Micromachining

Edited by Zdravko Stanimirović
and Ivanka Stanimirović

Published in London, United Kingdom

IntechOpen

Supporting open minds since 2005

Micromachining
http://dx.doi.org/10.5772/intechopen.75346
Edited by Zdravko Stanimirović and Ivanka Stanimirović

Contributors
Tatsuhiko Aizawa, Tadahiko Inohara, Elena Kralkina, Andrey Alexandrov, Polina Nekludova, Aleksandr Nikonov, Vladimir Pavlov, Konstantin Vavilin, Vadim Sologub, Vadim Odinokov, Tao Wu, Zhongli Zhang, Yushan Ni, Jinming Zhang, Can Wang, Xuedi Ren, Sung-Hua Wu, S Santosh Kumar, Ravindra Mukhiya, Wayne Hung, Mike Corliss

First published in London, United Kingdom, 2019 by IntechOpen
IntechOpen is the global imprint of INTECHOPEN LIMITED, registered in England and Wales, registration number: 11086078, 7th floor, 10 Lower Thames Street, London, EC3R 6AF, United Kingdom
Printed in Croatia

British Library Cataloguing-in-Publication Data
A catalogue record for this book is available from the British Library

Additional hard and PDF copies can be obtained from orders@intechopen.com

Micromachining
Edited by Zdravko Stanimirović and Ivanka Stanimirović
p. cm.
Print ISBN 978-1-78923-809-9
Online ISBN 978-1-78923-810-5
eBook (PDF) ISBN 978-1-83962-780-4

For EU product safety concerns:
IN TECH d.o.o., Prolaz Marije Krucifikse Kozulić 3, 51000 Rijeka, Croatia, info@intechopen.com or visit our website at intechopen.com.

Meet the editors

Zdravko I. Stanimirović has been active in research and development work for the last 24 years. He received his M.S. and Ph.D. degrees in electrical engineering from the Faculty of Electrical Engineering, University of Belgrade, the Republic of Serbia in 1999 and 2007, respectively. Dr. Z. Stanimirović is currently an associate research professor at the Telecommunications and Electronics Institute IRITEL A.D. Beograd. Over the years he participated in several scientific projects funded by the Ministry of Education, Science and Technological Development of the Republic of Serbia. Dr. Z. Stanimirović is the recipient of the IEEE Transactions on Components & Packaging Technologies best paper award. His current research interests include micro/nano electro-mechanical systems and micro- and nano-scale sensors.

Ivanka P. Stanimirović has been involved in research and development work for more than 20 years. Currently, she is an associate research professor at the Institute for Telecommunications and Electronics IRITEL A.D. Beograd. Dr. I. Stanimirović earned her M.S. and Ph.D. degrees in electrical engineering from the Faculty of Electrical Engineering, University of Belgrade, Republic of Serbia in 1999 and 2007, respectively. Over the years she has worked on several scientific projects funded by the Ministry of Education, Science and Technological Development of Republic of Serbia. She is the recipient of the IEEE Transactions on Components & Packaging Technologies best paper award. Her current research interests include micro- and nanoscale sensors and reliability issues in micro/nano electro mechanical systems.

Contents

Preface

One of the biggest challenges in today's world is the fabrication of 3D structures of a smaller and smaller size. The widening range of their application induces the necessity to micromachine diverse materials, many of them being difficult to machine. Nowadays, depending on the material to be machined, a wide variety of micromachining processes are available. The goal is to produce complex parts at enhanced material removal rates, with adequate surface integrities and dimensional accuracies in optimum production times.

To present their work in the field of micromachining, researchers from distant parts of the world have joined their efforts and contributed their ideas according to their interest and engagement. This book is the result of their hard work. In this book, you will have the opportunity to understand the concepts of micromachining of advanced materials. Surface texturing using pico- and femto-second laser micromachining is presented, as well as the silicon-based micromachining process for flexible electronics. You can learn about the CMOS compatible wet bulk micromachining process for MEMS applications and the physical process and plasma parameters in a radio frequency hybrid plasma system for thin-film production with ion assistance. Last but not least, study on the specific coefficient in the micromachining process and the multiscale simulation of influence of surface defects on nanoindentation using quasi-continuum method provides us with an insight in modelling and simulation of micromachining processes.

The editors would like to thank the authors for their contributions and efforts to present their work in the manner that allows both professionals and readers not involved in the immediate field to understand the topic. We would also like to express special appreciation to the IntechOpen team for their dedicated work in making this book possible.

Dr. Zdravko Stanimirović and Dr. Ivanka Stanimirović
Technology Department,
Institute for Telecommunications and Electronics IRITEL a.d. Beograd,
Belgrade, Republic of Serbia

Section 1

Micromachining Techniques

Chapter 1

Micromachining of Advanced Materials

Wayne N.P. Hung and Mike Corliss

Abstract

Market needs often require miniaturized products for portability, size/weight reduction while increasing product capacity. Utilizing additive manufacturing to achieve a complex and functional metallic part has attracted considerable interests in both industry and academia. However, the resulted rough surfaces and low tolerances of as-printed parts require additional steps for microstructure modification, physical and mechanical properties enhancement, and improvement of dimensional/form/surface to meet engineering specifications. Micromachining can (i) produce miniature components or microfeatures on a larger component, and (ii) enhance the quality of additively manufactured metallic components. This chapter suggests the necessary requirements for successful micromachining and cites the research studies on micromachining of metallic materials fabricated by either traditional route or additive technique. Micromachining by nontraditional techniques—e.g., ion/electron beam machining—are beyond the scope of this chapter. The chapter is organized as following: Section 1: Introduction; Section 2: Requirement for successful micromachining: cutting tools, tool coating, machine tools, tool offset measuring methods, minimum quantity lubrication, and size effect; Section 3: Effect of materials: material defects, ductile regime machining, crystalline orientation, residual stress, and microstructure; Section 4: Micromachining: research works from literature, process monitoring, and process parameters; Section 4.1: Micromilling; Section 4.2: Microdrilling; Section 4.3: Ultraprecision turning; Section 5: Summary; and References.

Keywords: micromilling, microdrilling, ultraprecision turning, minimum quantity lubrication, additive manufacturing

1. Introduction

Recent technological advancement and market need for product miniaturization demand suitable processes to mass produce three-dimensional (3D) microcomponents. Although microelectronic manufacturing techniques can produce two-dimensional (2D) microdevices using silicon and other semiconducting materials, silicon is neither robust enough for demanding engineering applications nor biocompatible for biomedical applications. Biocompatible materials and superalloys are traditionally fabricated in bulk quantity by forging, casting, or extrusion. The recent explosion of additive manufacturing innovations has led to several revolutionary fabrication methods of engineering devices. Powder bed fusion techniques using energy beams or binding polymers to consolidate powders in sequential layers are commonly used for metals. As in casting and welding, fabrication of a complex product by fusing re-solidified layers would introduce point, line, and volume defects in the part: dislocation entanglement, porosity, solidification shrinkage, microcrack, significant residual stress, anisotropy, rough surface finish, distortion, and undesirable microstructure are among key issues for metallic components fabricated by additively manufacturing route.

Micromachining techniques can be applied to successfully fabricate engineering components—either in meso or micro scales—from robust or biocompatible bulk materials. Micromachining is also among the key post processing techniques to enhance the quality of additively built metallic components [1–3]. This book chapter provides necessary requirements for micromachining, and cites research studies on micromachining of metallic materials fabricated by traditional or additive techniques.

2. Requirement for successful micromachining

To obtain the same surface speed as in macromachining, a machine tool must:

a. Be capable to rotate a workpiece or tool at high rotation speeds at 25,000 rpm or above,

b. Control spindle runout to submicron level,

c. Have very robust mechanical and thermal structure that does not affect by vibration or thermal drift, and

d. Have high resolution tool positioning and feeding mechanisms.

Success of micromachining depends on tool quality and precision of a machine tool. Machine spindle runout, tool concentricity and tool positioning accuracy must be in the neighborhood of 1/100 of a microtool diameter or less for successful operation. Tolerance stack up for spindle runout, tool eccentricity, and wandering of a microtool cause cyclic bending of a tool that leads to a catastrophic failure. At a low rotating speed, the displacement of a spindle can be monitored with a sensitive mechanical indicator. However, this option is not applicable for machines that operates at few thousands rpm or above. Other non-contact techniques using capacitance, magnetism, or light would be more appropriate. A laser beam can be focused on a rotating precision plug gage. The spindle displacement is then recorded on a computer for further analysis and is displayed in either frequency or time domain. Commercial laser systems can provide displacement reading to $\pm$10 nm resolution.

2.1 Size effect

The parameters for machining and tooling that are successfully applied in macromachining do not necessarily scale down linearly for micromachining. It is relatively easy to have a rigid turning or facing microtool, but it would require careful planning to maintain rigidity of a high aspect ratio micromill or a microdrill. Geometries of macroscale and microscale drilling/milling tools are the same: tool diameter, number of cutting flutes, point included angle for microdrill, helix angle, web thickness, clearance angle, flute length, shank diameter, and overall length. A careful selection of microtools must consider the intended machined features and highest possible tool stiffness. The two most important geometries that affect the microtool stiffness are the tool diameter and flute length assuming the number of flutes have been chosen. It can be shown that the torsional stiffness of a mill/drill is proportional to (tool diameter)^4 and (flute length)$^{-2}$. For a specific mill/drill tool dimension, we must adjust the milling/drilling strategy accordingly to avoid tool breakage.

If we select a drill diameter of 0.2 mm instead of 0.5 mm, then such 60% reduction of diameter will result in a reduction in torsional stiffness ΔE of:

$$\Delta E = \frac{(D_2)^4 - (D_1)^4}{(D_1)^4} = \frac{0.2^4 - 0.5^4}{0.5^4} = -97\% \tag{1}$$

Similarly, if we choose the flute length of 1.2 mm instead of 1.0 mm, this 20% change in flute length will lead to a decrease in torsional stiffness ΔE of:

$$\Delta E = \frac{(L_2)^{-2} - (L_1)^{-2}}{(L_1)^{-2}} = \frac{1.2^{-2} - 1.0^{-2}}{1.0^{-2}} = -30\% \tag{2}$$

Machining parameters that are successfully used in macromachining are not necessarily applicable for micromachining. A published literature recommends milling speed of 178 m/min and chip load of 0.1 mm/tooth for end milling 316L stainless steel using uncoated carbide tool.[1]

- Macromachining: to have the said surface speed for an Ø25.4 mm end mill, the required spindle speed is:

$$N = \frac{V}{\pi D} = \frac{178\ (\text{m/ min})}{\pi(\text{rad/rev}) \times 25.4\ (\text{mm})} \times 1000\ (\text{mm/m}) = 2230\ \text{rpm} \tag{3}$$

- Micromachining: using the same surface speed for an Ø0.1 mm micromill, the new spindle speed is:

$$N = \frac{V}{\pi D} = \frac{178\ (\text{m/ min})}{\pi(\text{rad/rev}) \times 0.1\ (\text{mm})} \times 1000\ (\text{mm/m}) = 555,600\ \text{rpm} \tag{4}$$

A machine tool with spindle speed exceeding 500,000 rpm is rare or simply not commercially available at this time. Applying the recommended macro chip load of 0.1 mm/tooth for an Ø0.1 mm micromill would break the fragile tool since the feed/tooth is as large as the microtool diameter.

2.2 Tool sharpness

The tool edge radius is critical in micromachining. If the depth of cut (or chip load) is too shallow, the tool simply plows the material and pushes it away elastically. This elastic material layer just springs back after the tool passing. If the depth of cut (or chip load) is substantial, then a chip is formed and a new machined surface is generated. Typical fine grain carbide tools are first sintered from submicron carbide particles in a cobalt matrix, and then ground and lapped to final

[1] Machinery's Handbook, 28 ed., Industrial Press, 2008.

geometry. Optimal edge radii of 1–4 μm are typically designed for sintered tools to balance edge sharpness and edge strength. Only single crystalline diamond tools can be ground and lapped to form edge radii within nanometer range.

The threshold depth has been investigated theoretically and verified experimentally by many researchers. It varies from 5 to 40% of the tool edge radius depending on the workpiece material and original rake angles. A depth of cut (or chip load), therefore, can be conservatively set to be 50% of the tool edge radius. When machining below this threshold, a microtool just rubs the surface and deforms it elastically during the first pass. When machining with depth of cut below the critical level, the material is then being plowed at negative effective rake angle. This results in high cutting force, high specific energy, fast tool wear, rough surface finish, and significant burr [4]. In subsequent passes when the cumulative depth is above the critical depth of cut, then a tool can remove materials as chips and the cycle repeats.

It is crucial to verify the tool edge radius before deciding on cutting parameters. Measuring of tool edge radius, however, is not trivial. A tool edge radius can be estimated from a scanning electron microscopic picture when the cutting edge is parallel to the electron beam axis [5], or scanning probe microscopic picture using a probe to scan the neighborhood of a cutting edge (**Figure 1**), or by scanning the edge on an optical microscope profiler in different views to reconstruct a 3D image of an tool edge before finding its radius.

2.3 Tool materials

Having the right microtool is essential for micromachining. A microtool that successfully drills through holes on a plastic printed circuit board does not necessarily be able to drill deep blind holes on titanium alloys. Understand the requirement and select the right microtool for each condition would save time, money, and frustration.

It has been theoretically derived and experimentally proven that the smaller the chip is, then the higher the required stress will be. Microcutting tools, therefore, have to be designed for higher stress with extreme geometrical constraints. When depth of cut is smaller than the average grain size of a workpiece, each grain with different orientation generates different stress on a cutting edge and eventually fatigues the tool.

Microtools as small as 25 μm are commercially available. Common tool materials are high speed steel (HSS), cermet, carbide, cubic boron nitride (CBN),

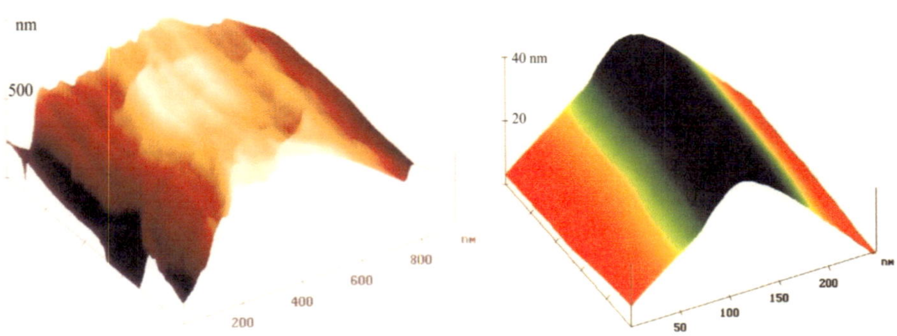

(a) (b)

Figure 1.
Tool edge radii of (a) 750 nm on a new polycrystalline diamond tool and (b) 10 nm on a new single crystalline diamond tool.

polycrystalline diamond (PCD), and single crystalline diamond (SCD). The HSS is not used in micromachining of metal since it does not have required hardness and strength to resist plastic deformation. A SCD tool is available for microturning, but not for microdrilling or micromilling. Carbide and cermet, having properties between HSS and diamond, are most suitable for microcutting tools. They are sintered from random abrasive grains in either cobalt or nickel binder with a small addition of molybdenum or chromium. A higher binder content increases the tool toughness and crack resistance, but reduces the bulk tool hardness. Having ultra-fine grain (submicron size) abrasives in a lesser amount of binder is the optimal solution since a tool with a submicron carbide grains can maintain a high hardness while improving its crack resistance against chattering, interrupted cut, or cyclic deflection due to spindle runout.

Microtool failure modes include shearing, chipping, and wear. To minimize shearing and catastrophic tool failure, a tool should be made from a high hardness substrate and its geometry would be suitable for micromachining, i.e., large included angle and sharp cutting edge. A tool with smaller than minimum included angle will be deformed and fractured in service.

Coating of microtool is still a technical challenge due to conflicting constraints for tool performance. Chemical or physical vapor deposition (CVD or PVD) techniques have been developed to coat cutting tools with mono/multiple layers of intermetallic or ceramic compounds (**Table 1**). Criteria for acceptable tool coating are numerous: uniformity of coating thickness, high hardness, high toughness, low friction, high wear resistance, surface smoothness, high chemical/diffusion resistance, and high temperature stability at a reasonable cost [6]. Although a coating thickness of 2–4 μm is acceptable for a macrotool, coating thickness on a microtool should be thinner, in the range 1–2 μm, to minimize fracture and peeling of the coating (**Figure 2**). Both CVD and PVD processes not only add the coating thickness to an edge radius, but they also increase the radius due to extra coating at a sharp corner. This is unfortunate since the thicker coating reduces the tool sharpness by enlarging the tool edge radius and causes an unfavorable plowing effect with negative effective rake angle. An uncoated microtool might perform satisfactorily, but the same machining parameter can be devastating to an over-coated microtool. Although the Calo destructive test can be used to measure coating thickness on a large object [8], it is more practical and convenient to measure coating thickness on an expensive microtool nondestructively.

Coating	Structure	Hardness (GPa)	Coefficient of friction	Coating thickness (μm)	Maximum temperature (°C)
TiN	Monolayer	24	0.55	1–5	600
TiCN	Gradient	37	0.20	1–4	400
TiAlCN	Gradient	28	0.30	1–4	500
TiAlN	Multilayer	28	0.60	1–4	700
AlTiN	Gradient	38	0.70	1–3	900
ZrN	Monolayer	20	0.40	1–4	550
CrN	Monolayer	18	0.30	1–4	700
Diamond like	Gradient	20	0.15	0.5–1.5	400
$AlTiN/Si_3N_4$	Nanocomposite	45	0.45	1–4	1200
$AlCrN/Si_3N_4$	Nanocomposite	42	0.35	1–5	1100

Table 1.
Commercial coating for microtools.

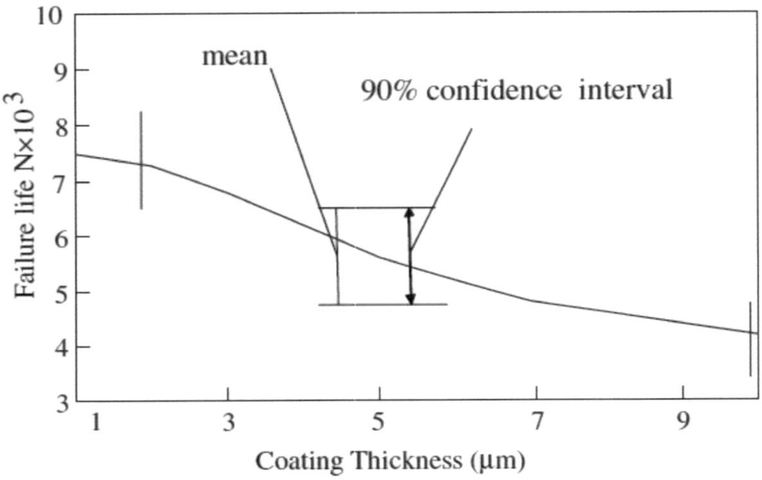

Figure 2.
Effect of tool coating thickness on tool life. TiC coated WC tool in interrupted cut [7].

Commercial instruments are available for coating thickness measurement using x-ray, magnetism, Eddy current, or ultrasound. A thin coating less than 1.5 μm following by an edge sharpening process would improve the tool performance, however, at the expense of higher tool cost. Published data indicate that micrograin carbide tools with 1.5 μm-TiN coating is the best for micromilling of H13 tool steel that has been hardened to 45 HRc.

2.4 Tool offset and positioning

Tool offset and tool positioning are crucial in micromilling and microdrilling since a high aspect ratio tool is small and extremely vulnerable. Selection of a suitable sensor for tool offsetting and tool positioning depends at least on following criteria:

- The sensor has better resolution compared to that of machine tool axis.
- The sensor should have a small working zone to cover a microtool.
- The sensor can perform at fast sampling rate for intended tool speed.

Contact techniques must be used with care for positioning a microtool. Common shop practices to find tool offsetting and positioning often damage a microtool or workpiece.

- A mechanical edge finder is adequate for most macromachining setup, but it is not suitable for micromachining especially with small and pliable part.
- Measuring resistance or current flow when a tool touching a conductive workpiece has been used with moderate success. A pulsed current might spark and damage the sharp edges of a microtool.
- An accelerometer can be mounted on either a workpiece or tool spindle housing. The difference in vibration signals indicates contact of tool and workpiece. The vibration signal, however, depends on the material of workpiece and tool, their surface roughness and detection threshold [9].

Non-contact techniques, although are generally more expensive, can provide a satisfactory accuracy and repeatability.

- Commercial laser displacement sensor with 20 μs sampling rate (50 kHz) would be sufficient for most cases. Submicron accuracy and be achieved, but the sensor's repeatability depends on the repeatability of multiple axes of a machine tool. Both lateral and axial tool offsets have been successfully used with this technique.

- Other non-contact techniques using magnetism, capacitance, ultrasound... could be used depending on the required accuracy and the workpiece materials.

2.5 Tool damage

Tool damage can be categorized by the relative size of the damage, ranging from submicron to hundreds of microns (**Table 2**). The tool failure mechanisms include damages due to mechanical, thermal, chemical effects, or adhesion.

Mechanical effect is the most common source of tool damage. Abrasive wear is caused by low speed sliding of hard particles from workpiece or tool against the cutting tool surface (**Figure 3**). At a high cutting speed and lack of sufficient coolant/lubricant, the high temperature at tool cutting edge accelerates the tool wear due to increasing rate of diffusion and/or chemical reaction at the coating layer and the substrate below. **Figures 5a** and **5b** compare the wear of PVD coated TiAlN/TiN layer on WC tool when machining at 180 m/min; the material contrast in scanning electron microscopy highlights the faster wear rate of the coating layer when machining 3D printed titanium alloy [10].

- Attrition wear is larger than abrasion wear. This happens when one for few grains of the tool are weakened at their grain boundaries and dislodged from the tool.

- Microchipping and chipping happen when larger chunks of tool being removed due to mechanical or thermal shocks upon loading and unloading (**Figure 4**). Machining at optimal parameters and rigid setup would reduce vibration, shock, and mechanical damage to a microtool. Chipping can occur due to high stress when machining at excessive cutting speed and feed [11]. Tool chipping also starts with microcracks due to chemical reaction among tool coating material, workpiece material, and coolant/lubricant.

Microtool damage	Damage size (μm)	Mechanism
Abrasion	<1	Mechanical, thermal
Attrition	1–3	Mechanical, thermal
Peeling	1–3	Mechanical, chemical
Microchipping	3–10	Mechanical, adhesion
Chipping	10–30	Mechanical
Fracture	>100	Mechanical

Table 2.
Categories of tool damage.

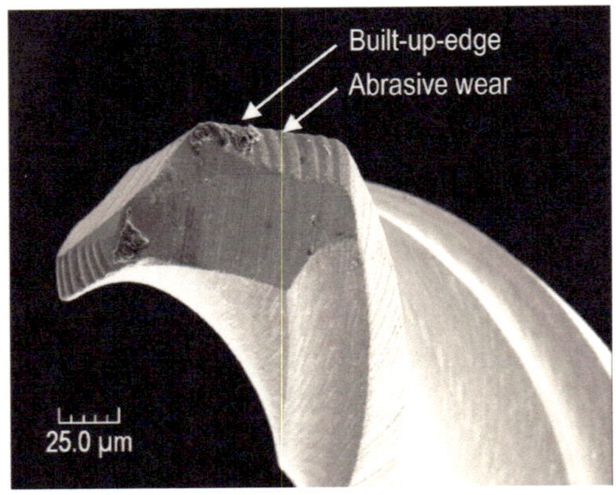

Figure 3.
Abrasive wear on a WC microdrill.

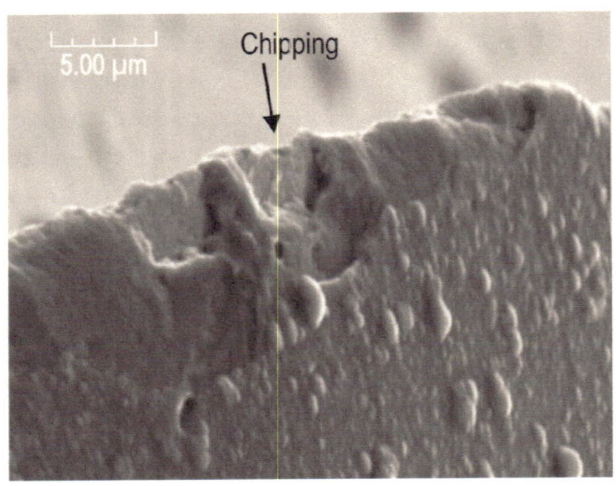

Figure 4.
Chipping of cutting edge. AlTiN coated micromilling tool.

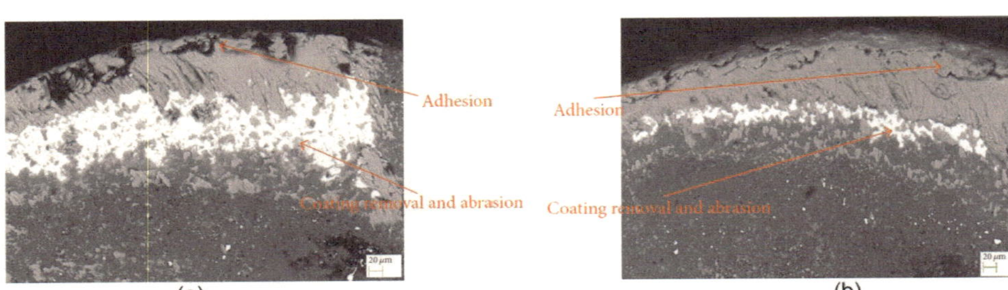

Figure 5.
Wear of coating tool after machining Ti 6Al 4V at 180 m/min on (a) wrought material and (b) selective laser melted material [10].

Thermal effect is the second cause of tool damage. A cutting tool edge is softened at high machining temperature, deformed plastically, and removed from the tool. Both high speed steel tool and carbide tool with high cobalt content are vulnerable to thermal damage. High temperature also promotes diffusion, i.e., atoms from the tool and workpiece move mutually across their interfaces, therefore degrade their properties and cause diffusion wear. Diamond with a carbon-rich matrix, or diamond-like coated tool, cannot be used with low-carbon ferrous alloy like steels or stainless steels since diamond carbonizes at temperature exceeding 600°C and diffuses to steel due to the steel's lower carbon content and its high affinity to carbon. A tool would extend its useful life by applying proper coolant to reduce thermal damage, or having a protective coating that blocks undesirable thermal diffusion from/to a tool surface. At higher cutting speed, the thermal/diffusion wear is the main tool wear mechanism. Combination of abrasive and thermal wear can be present when both high cutting speed and material hardness are combined. After machining at a high cutting speed of 180 m/min, severe coating tool wear (**Figure 5b**) is observed after cutting the harder selective-laser-melted titanium alloy compared to that when machining the same but softer extruded material (**Figure 5a**).

Chemical damage of a tool is due to chemical reaction between tool material and its environment like air, cutting fluid, or workpiece material. Tool oxidation is common when cutting in air at high speed. An oxidation reaction is accelerated with temperature, but can be eliminated when using inert gas to shield the cutting tool from surrounding oxygen. A chain reaction can also occur and further degrade a tool. For example, iron in steel is first oxidized at high cutting temperature to form iron oxide; this iron oxide then weakens the aluminum oxide coating of a tool and leads to peeling and chipping of the coating.

Adhesion tool damage happens when a built-up-edge (BUE) welds strongly to a tool surface and then breaks away with a minute amount of tool material. Some of the BUE deposits on the back of a chip, but some can be on the machined surface thus degrading the workpiece quality (**Figure 6**). When machining soft materials, a chip tends to adhere to the tool and grows in size (**Figure 7**). When such cumulative BUE is large and becomes unstable, it is removed with chip while shearing off part of the cutting tool due to the higher adhesion strength of BUE and tool interface than the inter-grain binding strength. Stainless steel, nickel and titanium alloys are known for causing adhesion wear on carbide microtools (**Figure 8**). Adhesion damage can be reduced by using proper lubricant to reduce friction between chip and tool, by coating tool with a smooth and low friction layer, by reducing tool edge radius, or by increasing cutting speed to raise tool surface temperature and soften BUE while reducing its weldability to tool surface.

A thick tool coating, although possessing a thicker diffusion barrier, can fail prematurely due to excessive shear stress at the interface (**Figure 9**). A 5 μm thick coating is common for large carbide insert, but 1–2 μm thin coating is recommended for microtools (**Figure 2**). Failure of microtools can happen due to combination of the above mechanisms. For example, peeling of tool coating might be due to coating defects or mechanical mechanism when a large gradient of stress exists across a

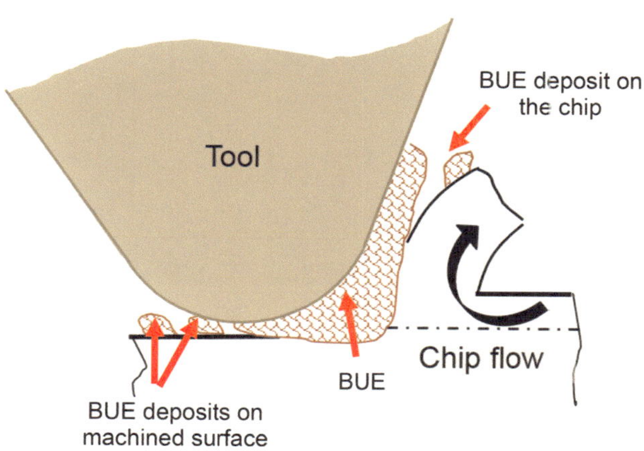

Figure 6.
Built-up-edge and its effects [12].

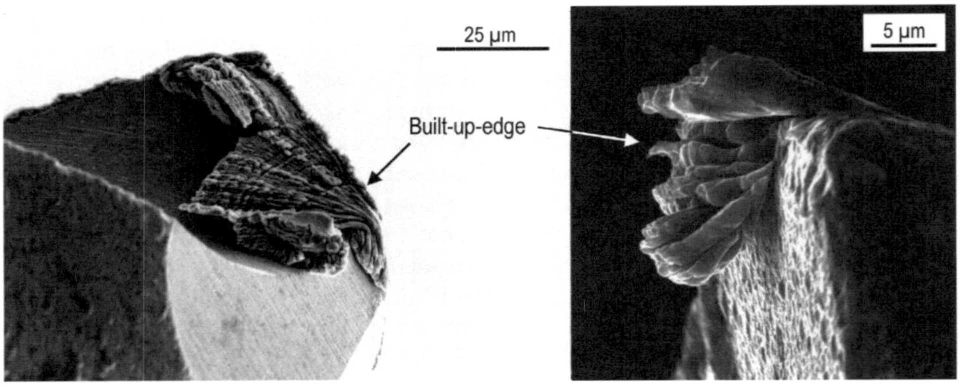

Figure 7.
Built-up-edge at the cutting edge on a microdrilling tool (left) and micromilling tool (right).

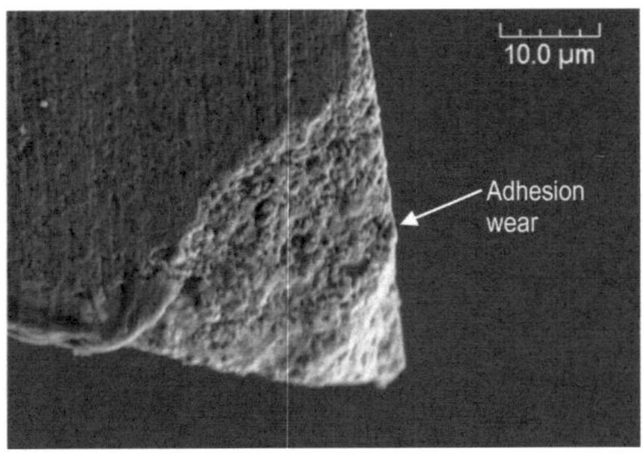

Figure 8.
Adhesion wear of a micromilling tool.

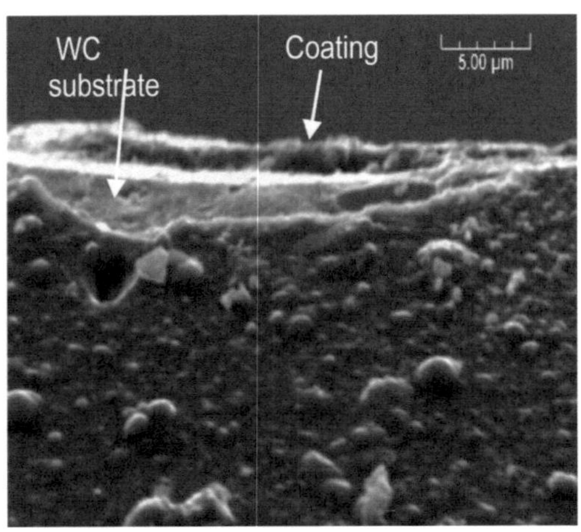

Figure 9.
Delamination of AlTiN/Si_3N_4 coating on a WC microdrilling tool.

thick coating layer; the loosen coating particles then rub and cause mechanical abrasive wear on a tool. Thermal mechanism may cause workpiece atoms to diffuse, weaken, and dislodge several tool grains as microchipping. The quenched and tempered 4140 steel fails tools by abrasion, the 304 stainless steel causes adhesion tool wear, while the nickel supper alloy Inconel 718 damages tools by all wear mechanisms [13]

2.6 Cutting fluid

Shearing of workpiece material and relative motion between tool and chip generate a significance of heat during machining. This thermal energy could change the microstructure, plastically deform the subsurface, degrade the part quality and wear a cutting tool quickly. Cutting fluids, either oil based for lubrication or water based for cooling, should be applied appropriately for effective micromachining while evacuating tiny chips from the machined surface.

- Dry machining. Although simple, dry machining is not appropriate since it neither reduce the heat, extend tool life, nor removing chips that may interfere with machining action.

- Flood cooling. A large amount of fluid can cover a tool and workpiece, but it is not effective since the bulk liquid cannot penetrate the air boundary layer surrounding a rapidly rotate micro drilling/milling tool-typically in the range of 30,000–120,000 rpm. Increasing the flood cooling pressure, as in jet cooling, simply deflect a fragile microtool and affect the machining quality.

- Minimum quantity lubrication (MQL). A mixture of oil and compressed air is very effective for micromachining when operating at high pressure above 4 bars (400 kPa, 60 psi). The micron-sized oil droplets can be propelled at high speed to penetrate the air boundary layer, adhere to workpiece/tool zones, spread out by surface tension to effectively cool and lubricate the tool/chip interface. Correct applications of MQL extend tool life while reducing burr as reported in published literature. Advanced MQL systems include additives (lignin, nano-sized diamond particle, graphene, etc.) can further enhance the effectiveness of MQL [14].

- Cryogenic cooling. Rapid freezing of most metals at liquid nitrogen temperature (−196°C) would embrittle the materials, reduce the required energy for machining and burr formation at the expense of tool wear [15]. This expensive technique, however, requires proper insulation of tooling and fixture surrounding a workpiece.

3. Effect of materials

Micromachining is often utilized to fabricate components for miniaturized sensors, medical, optical, and electronic devices. Common engineering materials for these applications include stainless steel, aluminum, titanium, copper, and tool steel for miniature molds and dies.

Workpiece materials must meet certain conditions for successful micromachining. Unlike in macromachining, a micromachining tool is subjected to fluctuating cutting force when encounters each grain since microtool size is comparable to material grain size. A microtool is more vulnerable to fatigue fracture and the resulted surface—if the tool survives—would be rough due to different springback protrusion from each grain due to different crystallographic orientations of the grains, and direction-dependent properties of the material. Homogenous workpiece materials with very fine and uniform grain sizes should be chosen for micromachining. Inclusions and large precipitates should be minimized to avoid damage to a fragile tool edge.

3.1 Ductile regime micro/nano machining

The concept of ductile-regime machining has been investigated since 1960s for amorphous brittle materials such as glasses. Silicon, germanium, and glasses have become strategic materials that are widely used to fabricate intricate components in microelectronics, optics, defense industries, and recently as micro optical-electrical-mechanical systems. Silicon and other brittle materials are known for their low machinability unless they are machined in the ductile-regime conditions. When utilized at the optimal machining conditions, only minimum effort is required for the subsequent etching, grinding, or polishing to remove the damaged subsurface. This section summarizes the theory and provides practical guidance for ductile-regime machining.

The mechanism of ductile-regime machining has been studied by many researchers. Using fracture mechanics approach, it can be shown that there is a threshold below which ductile regime prevailed:

$$d_c = \frac{\text{plastic flow energy}}{\text{fracture energy}} = A\left(\frac{E}{H}\right)\left(\frac{K_c}{H}\right)^2 \tag{5}$$

where d_c: critical depth of cut (m); A: constant. A = 0.15 for microscratching, A = 0.60 for micromachining; E: Young modulus (Pa); K_C: surface fracture toughness (Pa m$^{0.5}$); H: surface microhardness (Pa).

A shallow depth of cut, therefore, would energetically promote plastic flow rather than brittle fracture in the substrate and the chips. **Table 3** tabulates properties of some brittle materials and their experimental critical depths of cut.

The constant A in Eq. (5) varies in the range 0.1–0.6 due to measuring uncertainty of surface toughness K_c, elastic modulus E, and microhardness H in a testing environment. These properties depend on crystalline orientation of the materials, surface conditions, and tool geometry. An example of ductile regime machining on single crystal silicon wafer is shown in **Figure 10**.

- The critical resolved shear stress, on a crystalline plane due to the cutting action, is directly proportional to the Schmid factor $(\cos\lambda)(\cos\phi)$, where ϕ and λ are the orientations of the slip plane and slip direction. An ideal ductile mode machining would happen when the cutting shear stress is parallel to both the slip plane and the slip direction, otherwise a pseudo ductile mode with micro cleavages occurs. True ductile-regime machining happens only along certain crystalline orientation, but brittle machining occurs at other crystalline orientation. This explains why micromachining a crystalline specimen at the

Materials	Young modulus (GPa)	Fracture toughness (MPa m$^{0.5}$)	Knoop hardness (GPa)	Critical depth of cut (μm)
α-Al_2O_3	275–393	3.85–5.90	19.6–20.1	1.0
SiC	382–475	2.50–3.50	24.5–25.0	0.2
Silicon	168	0.6	10	0.5

Table 3.
Properties of selected brittle materials.

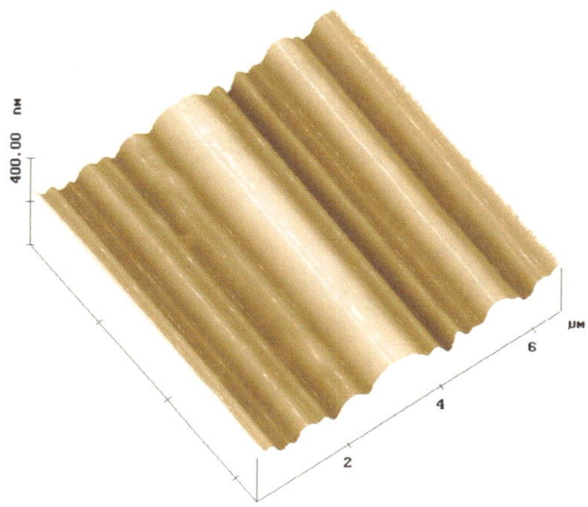

Figure 10.
Perfect ductile regime machining of (001) silicon [16].

same speed, depth of cut, and coolant produces ductile machined surfaces in one direction but brittle machined surfaces on others.

- Cutting fluid changes the surface properties of materials (K_c, E, and H) and affects conditions for ductile regime micromachining. When micromachining the (100) plane of germanium using a single crystalline diamond tool, the critical depth of cut changes from 0.13 μm with distill water as cutting fluid to 0.29 μm in dry machining.

- Tool geometry also affects the results. Plowing and fracture of material occurs when depth of cut is less than approximately half of the tool cutting edge radius. Tool with negative top rake angle is usually utilized since a negative rake causes compressive zone in the workpiece ahead and below the tool and suppresses microcrack formation.

3.2 Additively manufactured metals

The evolution of 3D printing allows metallic parts to be printed in different methods.

- Power beam fusion. Laser or electron beam are used to melt either metal wire or powder particles to form layers, then fuse these layers to form a complex shaped part.

- Material jetting. Metal powder is fed at the energy beam focused point to melt and form a part.

- Jet binding. An organic binding material is sprayed and bind metal powder in layers. The "green" part is then sintered to form the final shape.

Using any of the above techniques, an as-built part:

i. Is warped due to high thermal induced residual stress, or non-uniform shrinkage during sintering,

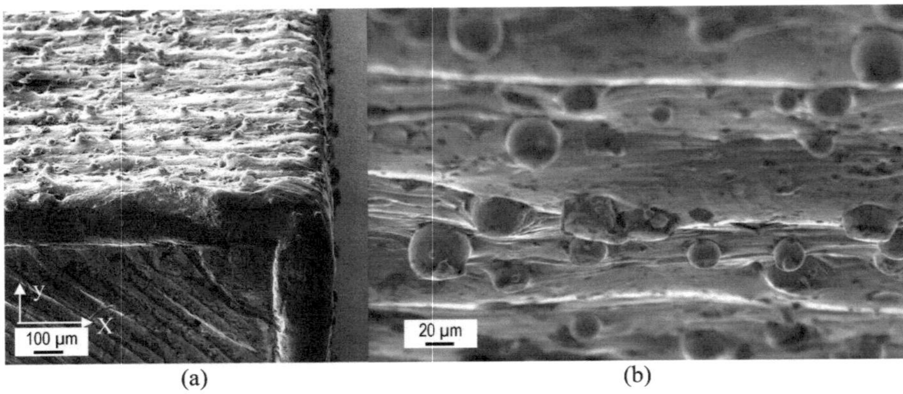

Figure 11.
Surface topography of an Inconel 718 block after selective laser melting. (a) Oblique view, and (b) viewing along the building z-axis.

ii. Has very rough surface (typically 15–20 μm Ra), and

iii. Is filled with surface defects (e.g., microcrack, shrinkage cavity, partially welded powder particles, etc.) as shown in **Figure 11**, and volume defects (e.g., porosity, inclusion, shrinkage cavity, etc.).

Post processing of the as-built metal parts must be performed so that they can meet required engineering criteria for surface finish or dimensional/form tolerances. Micromachining is the most effective post processing technique to control the surface and dimension of local areas of additively manufactured metals due to its high removal rate and well-established computer numerical controlled (CNC) industry.

4. Micromachining

Micromachining refers to removal of material subtractively in micron scale. The process can be done by (i) conventional processes, i.e., removing material mechanically with hard tools in contact with a workpiece and removing minute amount of material as chips, or (ii) non-conventional processes, i.e., removing material by other physical mechanisms such as optical, thermal, chemical, electrical, or combinations of these. The following section focuses only on the three major conventional techniques, namely micromilling, microdrilling, and microturning of advanced materials.

4.1 Micromilling

Micromilling is among the most versatile microfabrication processes. Although alternative nontraditional processes to produce microfeatures (e.g., laser micromachining, electrical discharge micromachining, electrochemical micromachining, chemical microetching, electron/ion beam micromachining) are available, these processes are either cost prohibitive, or inferior when comparing resulted surface and subsurface integrity, anisotropic aspect ratio, material removal rate, or feature quality. Successful micromilling requires new tool geometry, tool material, machining parameters, and machining skills. It is technically incorrect and costly to perform micromilling by just scaling down a milling cutter, or parameters from macroscale milling.

Tool material. Carbide tools should be sintered from fine grains, and precisely ground to obtain a micron-level cutting edge radius.

Milling direction. Down milling is the preferred mode since a micromill will engages a workpiece and removes a wedge shape chip with decreasing chip thickness. In contrast, a tool in up milling would rub on the workpiece until the effective chip thickness is greater than 0.5 cutting edge radius. Down milling also produces less amount of burr.

Lubrication. Minimum quantity lubrication should be used with all micromachining. A nozzle should be as closed as possible and is positioned to let the cutting flute pulls the mist into the cutting zone. Tool and workpiece should be arranged to avoid stagnant zone, or being blocked or interfered by chips [17].

Tool vibration. Avoid unnecessary disengaging then engaging of microtool and workpiece in a milling program. Vibration and bending of a microtool at starting and ending could fatigue and shorten tool life of a microtool (**Figure 12**).

Commercial micromills are available for diameter of 25 μm and above. Optional geometries include:

- Flute lengths: standard or extended length (10–80% longer)
- Number of flutes: 2, 3, or 4
- Helix angles: 25°, 30°, 50°
- End configuration: hemisphere or flat.

As mentioned above, the BUEs on a cutting tool surface can break and deposit on a machined surface as shown in **Figure 13**. Measuring area surface finish Sa would combine the roughness contributed from milling parameters (e.g., speed, feed), tool and machine condition (e.g., vibration), and surface irregularities (e.g., pore and BUE). The BUEs is reduced when an optimal condition of MQL is used, low chip load, and higher speed. In a study of micromilling 316L stainless steel, the amount of BUE is significantly diminished when cutting at speed higher than

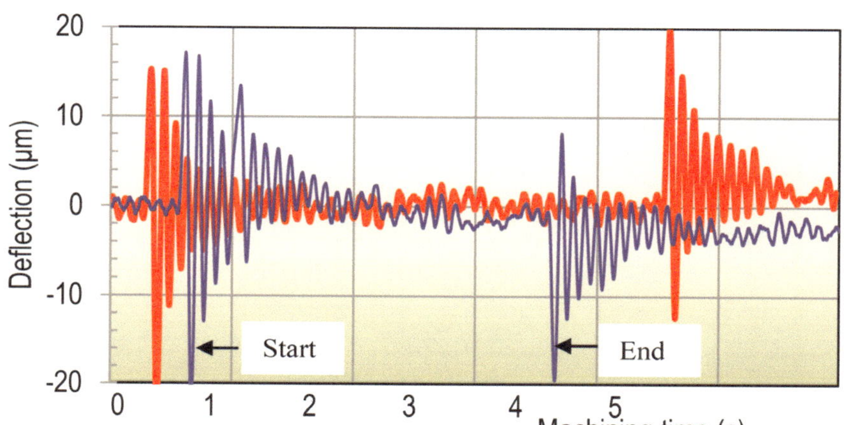

Figure 12.
Vibration of a micromill when engaging and disengaging a workpiece. Carbide mill Ø1 mm, 316L stainless steel, 25,000 rpm, 10 μm/tooth feed, 0.348 mm axial depth, 0.558 mm radial depth.

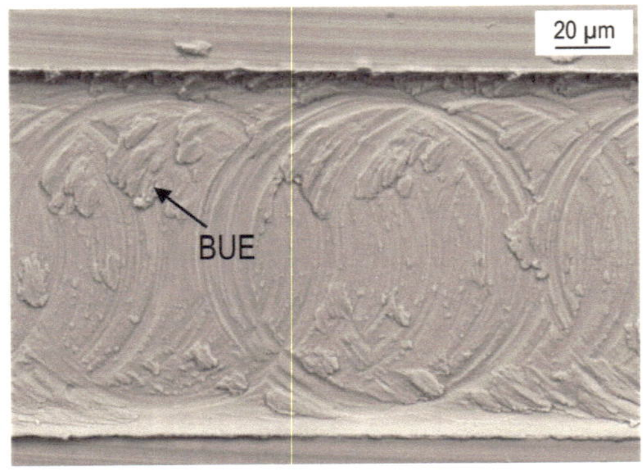

Figure 13.
Built-up-edges deposit on machined Ti 6Al 4V surface. Micromilling at 9.6 m/min, 0.1 μm/tooth, 10 μm axial depth. Dry [18].

30 m/min (**Figure 14**). Perhaps the high temperature at high cutting speed improves the material plasticity and reduces the weldability of BUEs on cutting tool tip.

Lack of BUEs on coated tool also results in better surface finish of micromilled channels on 304/316L stainless steels, NiTi alloy, A36 low carbon steel, 6061-T6 aluminum, and Ti-6Al-4V titanium alloy. The AlTiN coating effectively increases tool life while reducing burrs significantly when micromilling 304 stainless steel. The incompatibility of the coating on specific material workpiece prevent BUE formation, therefore, shear the materials as chips rather than deforming it as burr (**Figure 15b**). The effect of micromilling mode is also shown when up-milling tends to generate more burr than down-milling (**Figure 15a**).

The theoretical surface finish of machined surface after milling with a flat-end tool can be shown to be:

$$R_a = \frac{5}{18} f_t tan\alpha \tag{6}$$

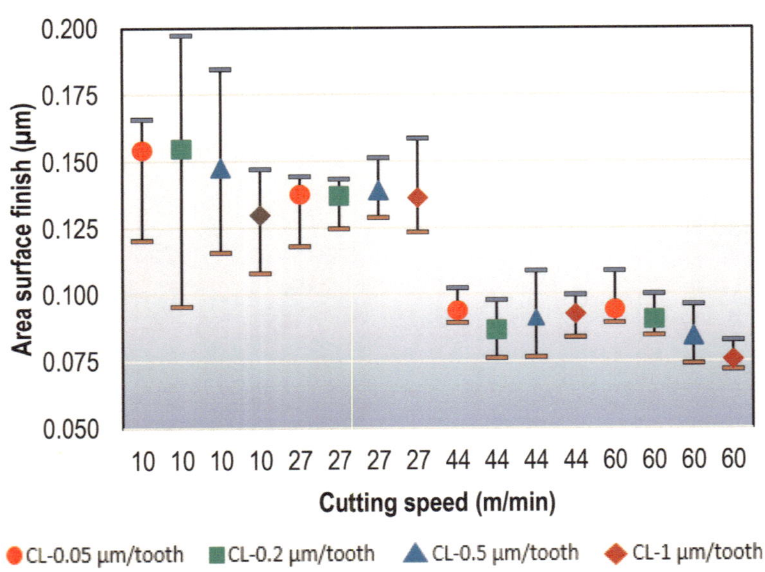

Figure 14.
Effect of cutting speed and chip load on area surface roughness Sa. Each range plot shows the maximum, minimum, and average of 15 measurements. Micromilling 316L stainless steel [12].

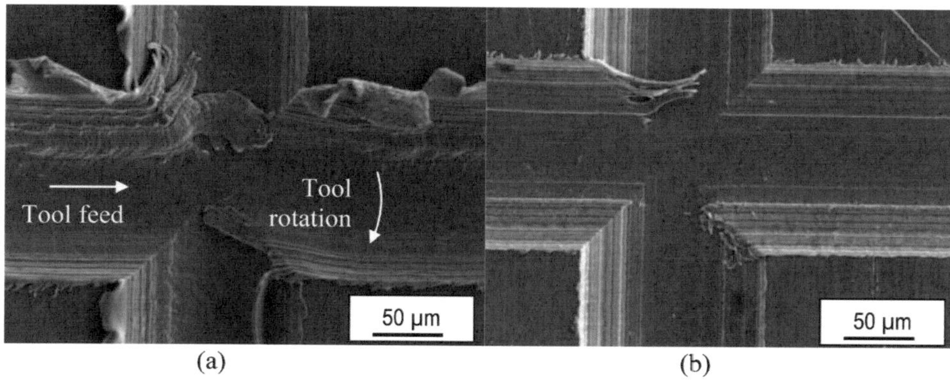

Figure 15.
Effect of tool coating on resulting burrs: (a) uncoated ϕ152 μm tool, milling 304 stainless steel, 24 m/min, 0.1 μm/tooth, MQL; and (b) AlTiN coated ϕ198 μm tool, milling 304 stainless steel, 24 m/min, 0.1 μm/tooth, MQL [19].

And that for a ball-end milling tool is:

$$R_a = 0.2423\frac{f_t^2}{D}. \tag{7}$$

Rearranging Eq. (7) to have:

$$D \cdot R_a = 0.2423f_t^2 \tag{8}$$

where R_a: average surface finish (μm); f_t: chip load (μm/tooth); D: diameter of ball-end milling tool (μm); α: concavity angle, or end-cutting-edge angle (°).

Both Eqs. (6) and (8) predict the dependent of chip load on surface finish Ra. When plotting Eq. (8) using experimental data from different tool diameters and different chip loads, then Eq. (8) is confirmed with data in macromilling when chip load >100 μm but not with smaller chip load for micromilling (**Figure 16**). The reason for a higher surface finish is the intermittent BUEs, although small, smear and degrade the machined surface.

Similar experimental results is shown in **Figure 17** for micromilling of additively manufactured Ti 6Al 4V alloys. At a very low chip load of 0.1 μm/tooth, the presence of significant BUEs on machined surface degrades the surface quality as indicated by high surface roughness R_a. The surface improves at higher chip loads, by reduction of BUEs, but is gradually increased with chip load as predicted by Eq. (6). Similar surface roughness result was reported for machining 7075 aluminum alloy [21].

Post processing by micromachining of additively manufactured metals have been investigated. Few studies have compared machinability of selected metals produced by conventional route (e.g., casting, extrusion, rolling, etc.) and by additive manufacturing route (e.g., powder bed fusion, direct energy deposition, etc.). Limited machining investigations on Inconel 625, Inconel 718, Ti 6Al 4V, H13 tool steel, Ti 48Al 2Nb 2Cr alloy, 17Cr 4Ni stainless steel, and 316L stainless steel have concluded that the AM metals in general have lower machinability compared to the conventional metals.

Machinability is affected by microstructural changes in a material. A quick comparison between the microstructure of extruded and SLM'ed Inconel 718 shows the contrast of the same materials after different manufacturing routes. Uniform

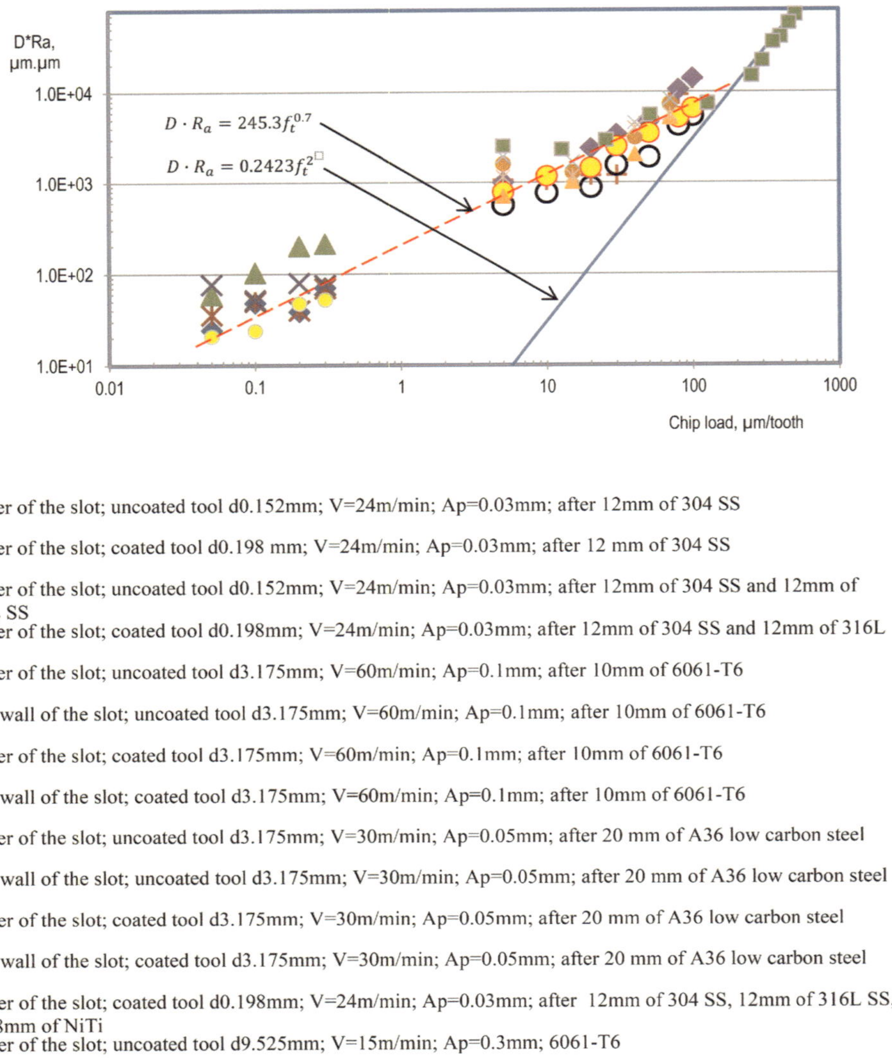

Figure 16.
Average surface finish at center of milled microchannels. Ball-end milling tools ϕ152–9525 µm, workpiece materials 6061-T6, A36 steel, NiTi, 304/316L stainless steels, in MQL condition [19].

microstructure of the extruded specimen is expected (**Figure 18**). The mechanical properties of extruded specimen could be slightly different in longitudinal and transversal directions due to preferred grain orientation along the extrusion direction. In contrast, the fast heating and cooling rate of SLM'ed Inconel 718 creates alternate layers along the laser paths and hard particles in the microstructure (**Figure 19**). The obvious changes in microstructure result in different mechanical properties, therefore, affecting machinability. Fast cooling in SLM'ed Inconel 718 also forms brittle Laves particles and traps pores near an edge (**Figure 20a**). During machining, some particles are broken, smeared along the tool path and probably chipped the cutting tool. Porosity is unavoidable in AM metals. The burr on top of those micron-size pores after micromachining is difficult to remove mechanically (**Figure 20b**). Although hot isostatic pressing (HIPping) can eliminate porosity, some materials (e.g., Inconel 718) are aged at HIPping temperature; the resulted precipitates increase the material strength but reduce its machinability (**Figure 21**).

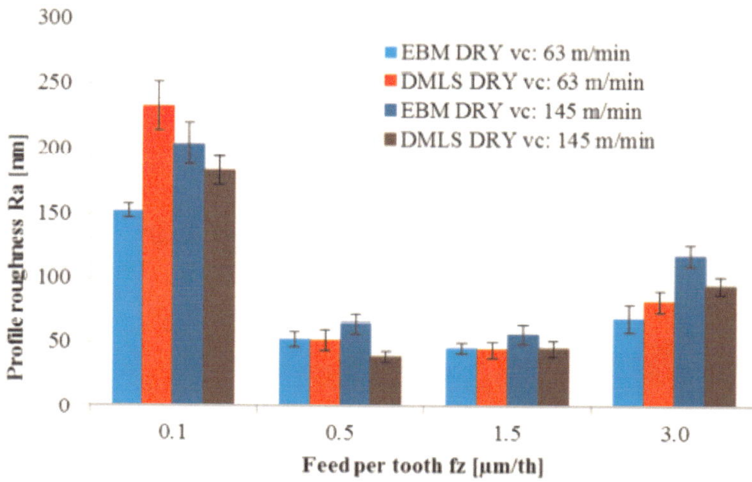

Figure 17.
Effect of chip load and cutting speed on surface finish. Micromilling of Ti 6Al 4V fabricated by electron beam melting, and direct metal laser sintering [20].

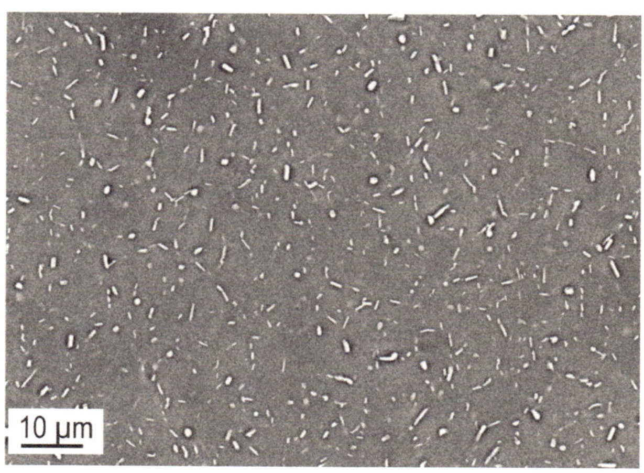

Figure 18.
Uniform microstructure of extruded Inconel 718. Viewing along the extruding direction.

Conflicting literature data are probably due to different process parameters for AM metals and different scanning strategies. For example, the precipitation hardenable 17Cr 4Ni stainless steel was reported to have near fully martensitic structure and high yield strength than the cast alloy, but the opposite conclusion was found in another study.

Advanced cutting fluids and techniques have been applied for micromilling of advanced materials. Poor results are reported when using a pressurized jet to flood cool during micromilling of Ti 6Al 4V. The fragile tool (200 µm diameter, AlTiN coated, 1.25 µm/tooth feed, 30,000 rpm rotating speed, 20 µm depth of cut) being deflected and vibrated under pressurized jet, generates large burr and rough surface (**Figure 22a**), while cutting a large slot width (**Figure 22b**). Applying MQL with flow in the feeding direction solves these problems. When using MQL at high air pressure above 5.5 bar, a nozzle with rough internal surface breaks the lubricant into smaller droplets and effectively improves tool life of micromilling tools [24].

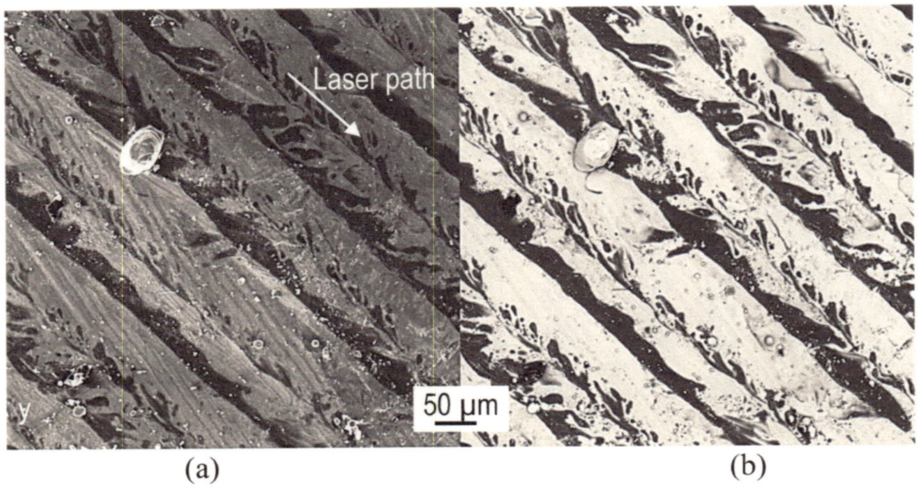

Figure 19.
Microstructure of SLM'ed Inconel 718 by scanning electron microscopy in (a) secondary electron mode, and (b) back scattered electron mode. Notice the different layers across the laser scanning paths.

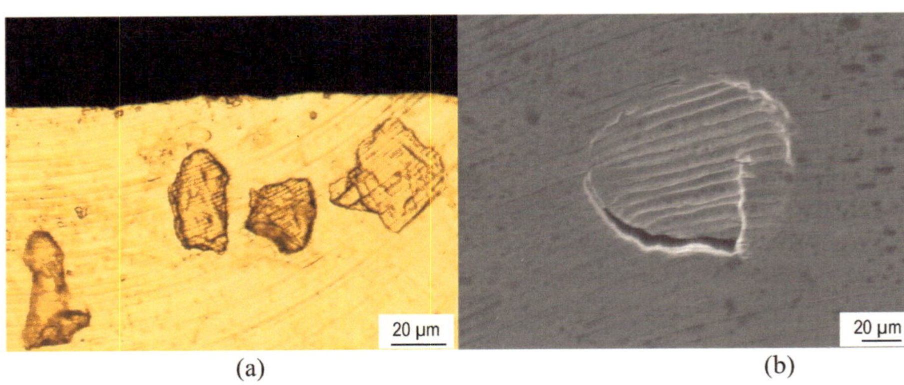

Figure 20.
Irregularities in selective laser Inconel 718. (a) Sheared Laves particles after micromilling, (b) A machined micropore with burr.

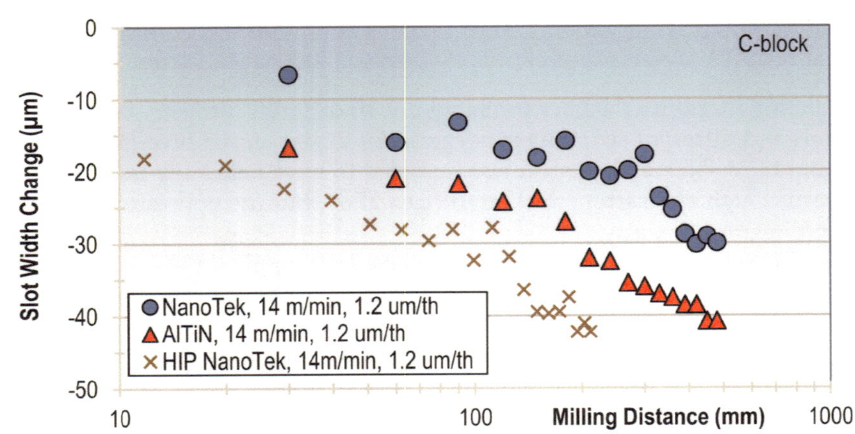

Figure 21.
Micromilling of SLM'ed Inconel 718. Change of slot width with milling distance. Minimum quantity lubrication, 50 µm depth of cut [22].

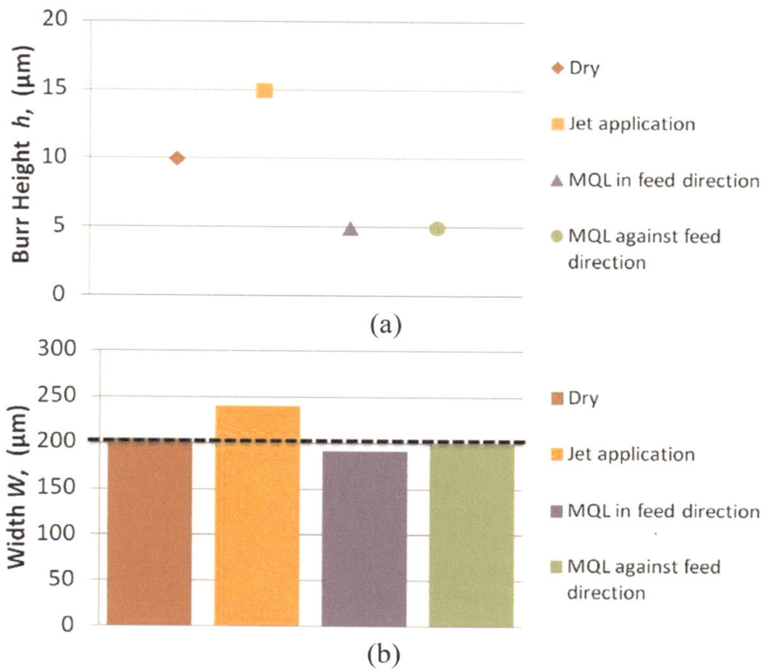

Figure 22.
Effect if different cooling methods on (a) burr formation, and (b) resulted slot width. Micromilling of Ti 6Al 4V [23].

Effort was made to produce an environmentally friendly cutting fluid while enhancing the cutting fluid performance. Lignin is a biodegradable product from wood. It can be mixed in alcohol then TRIM water soluble as cutting fluid. Micromilling tests are performed with 396 μm diameter milling tool on 6061 aluminum and 1018 steel. The optimal concentration of 0.015% lignin seems to provide the best lubricating and cooling effects when reducing the cutting forces on both materials (**Figure 23**).

Cryogenic precooling can be used to reduce BUE formation and its effect on part quality. Liquid nitrogen, dispensed in front of a microtool in a micromilling test on Inconel 718. The tool (760 μm diameter, AlCrN coated) is used at 48 m/min speed, 1.25–5 μm/flute chip load, 0.1–0.2 mm depth of cut for a constant distance of 120 mm. The cryogenic condition embrittles the Inconel material so it can be micromilled in brittle mode with minimum plastically deformed burr and BUEs. The result is the low surface roughness at different chip loads (**Figure 24a**) and depth of cuts (**Figure 24b**). However, the brittle chip debris are abundant and might interfere with subsequent machining passes. No chip debris is seen when MQL is used.

The positive results of cryogenic cooling on Inconel 718, however, was not confirmed in a similar study on micromilling of electron beam melted Ti 6Al 4V. The 300 μm cutting tools were utilized at 63–145 m/min speed, 0.1–3.0 μm/tooth chip load, and 30 μm axial depth of cut. The temperature of the workpiece was at -155 ± 5°C with liquid nitrogen. No brittle chip debris are seen, ductile burr are visible (**Figure 25a**), and the surface profile (R_a finish) are similar for dry, MQL, and cryogenic conditions. The machined surface is work-hardened after micromilling at cryogenic condition as shown in the nanohardness results below the surface (**Figure 25b**). This implies the material is not truly embrittled as planned.

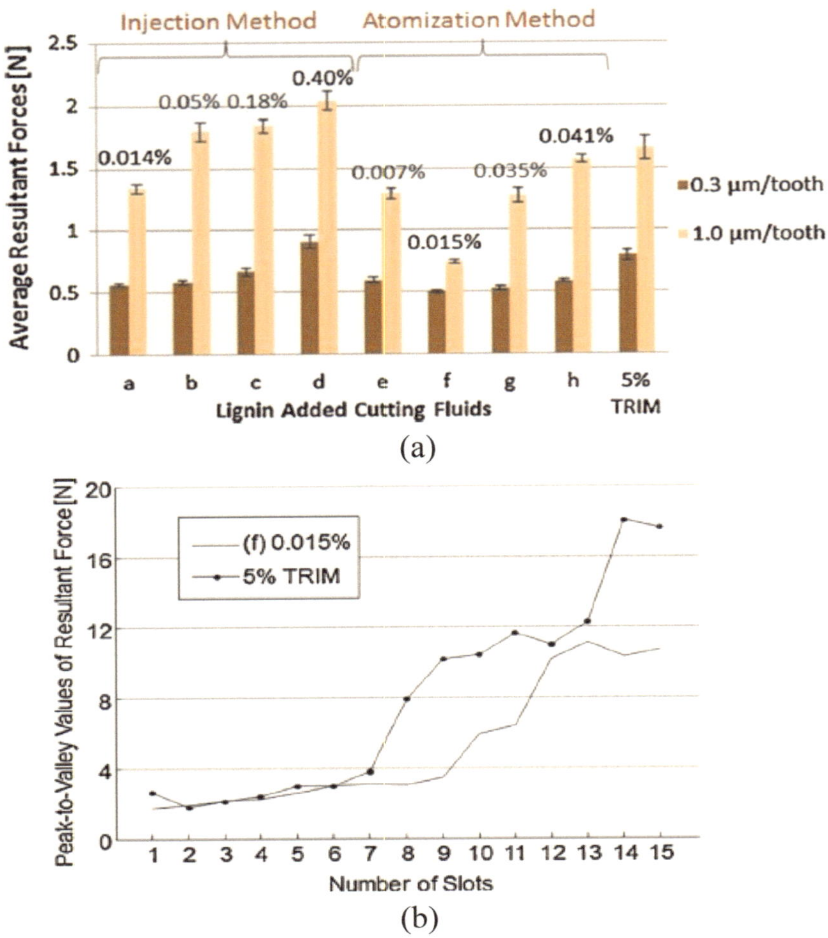

Figure 23.
Effect of atomized cutting fluid with lignin on micromilling of (a) 6061 aluminum and (b) 1018 steel [25].

Another effective additive in MQL is nano-sized diamond particles. This application, applied to microdrilling, will be presented in the next section.

4.2 Microdrilling

Microdrilling is a more complex operation comparing to turning or milling. Chip removal and effectively supplying of cutting fluid are easy with the latter, but not with microdrilling due to extremely limited space around a microdrill.

Tool material. As with a micromill, a carbide microdrill should be sintered from fine grains, and ground to small cutting edge radius.

Hole quality. Spindle runout, tool eccentricity, and wandering of a microdrill cause cyclic bending of a tool and could lead to a catastrophic failure. To control drill wandering, precision pre-drilling of a center hole can be tried, or the workpiece surface must be ground to minimize deflection of a slender drill when starting on an irregular surface.

Micromist with fixed nozzle pointing to the drill tip and making an angle of 60–70° with the tool axis is recommended. This way, the chip is blown away after each pecking cycle and the microdrill is re-lubricated before re-entering into the hole.

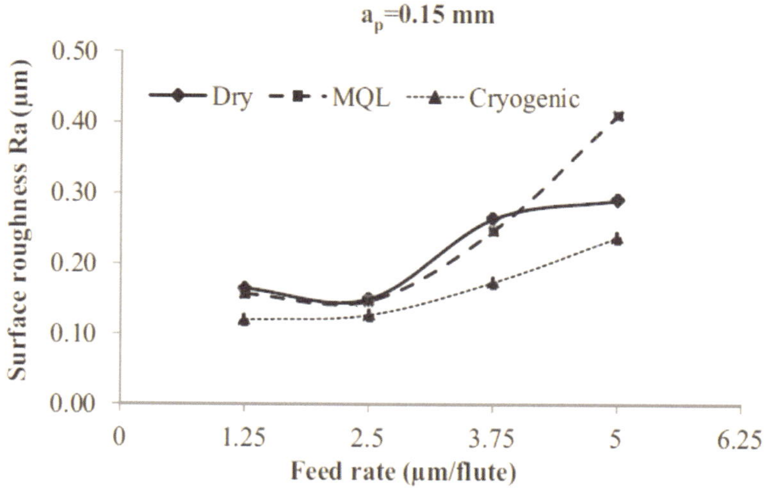

(a)

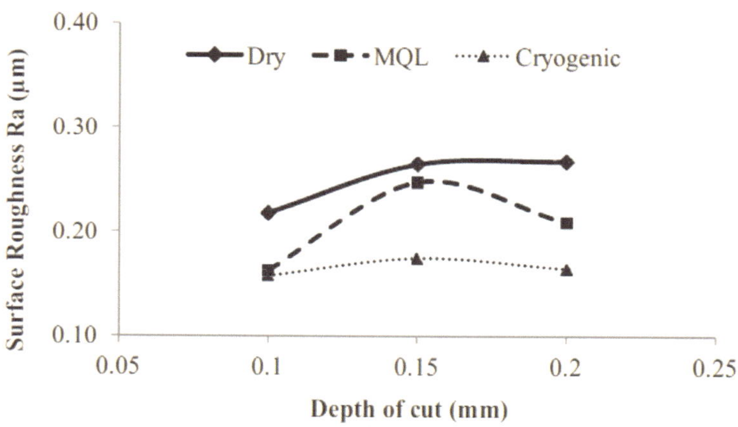

(b)

Figure 24.
Effect of different cooling strategies on micromilling of Inconel 718 (a) effect of chip loads, and (b) effect of axial cutting depth (Ucun, 2014).

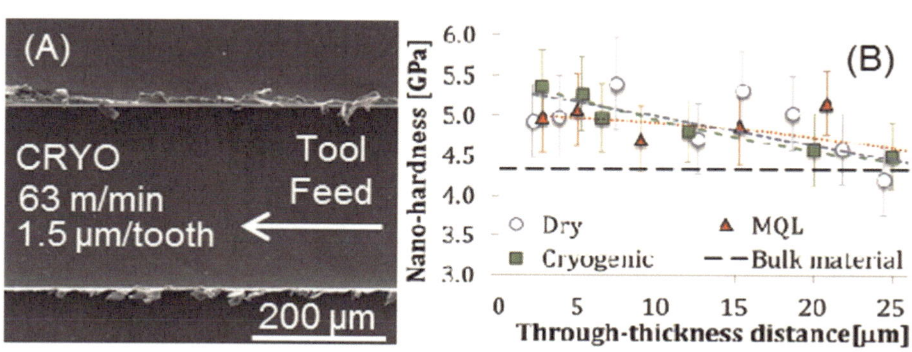

Figure 25.
Effect of cryogenic cooling on micromilling of electron beam melted Ti 6Al 4V [26].

High aspect ratio. Pecking is essential for microhole drilling since chips have to be extracted and cutting fluid must penetrate into a small and deep microhole. The pecking depth can be substantial in the beginning, but it must be gradually reduced when drilling at deeper depths. One can start with an initial pecking depth of (2*drill diameter) and gradually reduce it to (0.5*diameter) at the hole depth of (10*diameter). It is convenient to program pecking cycles in microdrilling following the equations below.

$$\frac{P}{D} = \frac{1}{9}(-1.5R + 19.5) \quad \text{for } R \leq 10 \tag{9}$$

$$\frac{P}{D} = 0.5 \quad \text{for } R > 10 \tag{10}$$

where P: incremental pecking depth (mm); D: drill diameter (mm); R: progressive drill aspect ratio = current hole depth/drill diameter.

Although application of MQL in micromachining is necessary, other researchers have found ways to improve its effectiveness. Nanoparticles are mixed in MQL oil to improve its performance in microdrilling. The nano-sized particles of CNT-C60, TiO_2, Al_2O_3, MoS_2, and diamond would increase the thermal conductivity of the fluid thus prolong the tool life while reducing burr. In an experimental microdrilling study, uncoated microdrills of 100–500 μm diameters are used at 10–15 mm/min, 30,000–60,000 rpm spindle speeds, while varying the concentration of 30 nm diamond particles in the range 0–4%. The MQL system is used at 3-bar air pressure to drill a constant 0.3 mm depth. The optimal conditions for low torque and thrust force are experimentally obtained to be 60,000 rpm and 2% concentration.

Inspection of a microdrilled and tapped hole would be difficult. Destructive technique by sectioning a part is time consuming, expensive, and error prone since the internal features might be distorted by releasing of residual stress. X-ray computed tomography (CT) has been utilized to evaluated drilled and tapped Ti 6Al 4V dental implant fabricated by direct metal laser sintering. The best microdrilled hole quality—cylindricity and perpendicularity—is achieved at the lowest testing drill speed of 60 m/min and the lowest chip load of 10 μm/flute. **Figure 26a** shows the sectional view of the CT image of a drilled and tapped dental implant. Detailed observation and measurement can then be performed (**Figure 26b**).

Finite element modeling of a microdrill was done to find the limiting drilling parameters that would fracture a microdrill catastrophically. It was assumed that

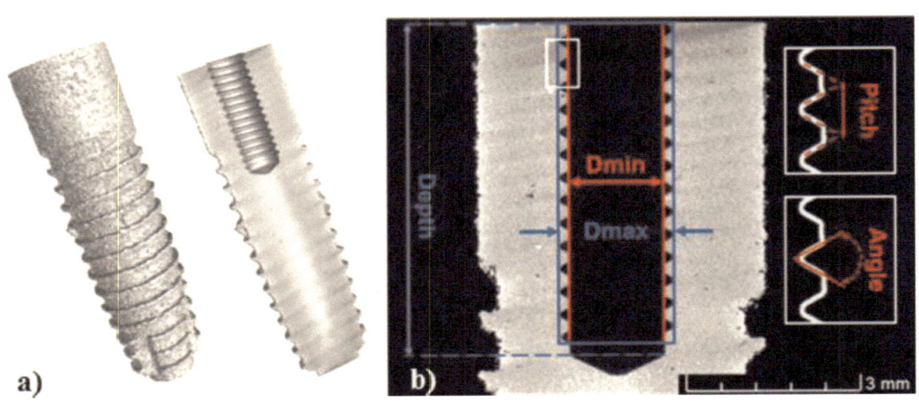

Figure 26.
Inspection of drilled and tapped microhole by X-ray computed tomography. Ti 6Al 4V dental implants fabricated by direct metal laser sintering [27].

the drill shank is rigid, axial thrust force and torsional torque can fail a drill in either buckling mode or torsional mode. Transverse shear strength in three-point bending test of a sintered carbide tool is estimated to be half of the material tensile strength. This study used commercially pure (CP) titanium, 316L stainless steel, 6061-T6 aluminum, PEEK plastic, and Nitinol (51 wt% Ni 49% Ti) shape memory alloy. Some surfaces were faced milled on a milling machine, others were milled, hand ground and then finish polished with 1 μm diamond paste. Microdrilling was performed with Ø100–150 μm drill diameter, 135° point angle, 30° rake angle, 40–44° helix angle, 2 flutes, and 1.50–3.50 mm flute length. Some were coated with AlTiN or $AlTiN/Si_3N_4$ under the trade name Nanotek.

The classical Taylor's equation has been applied for macromachining, micromachining, and is used for microdrilling to show the effects of chip load and tool coating. For the same cutting speed of 20 m/min and comparable drilling distance of about 35 mm, the CP titanium can be microdrilled 400% faster than 316L stainless steel since the chip load for the former is 0.1 μm and that for the latter is 0.02 μm. Also, AlTiN coated drills improve tool life by at least 122%. This drilling operation is stopped after drilling all possible holes on the test blocks (**Figures 27** and **28**). Negligible tool wear are observed when drilling 6061-T6 aluminum and PEEK plastic, therefore, no modeling is necessary.

Burr, hole size, hole position, and work hardening around a drilled hole contribute to the hole quality. During the initial engagement of drill and workpiece surface, a slight lateral motion of the drill chisel edge is sufficient to bend and misguide the slender microdrill. Drill wandering refers to the deviation of a drilled hole from its intended position. A machined surface is rough enough to cause wandering of microdrills. A sloped ridge on a rough surface bends and deviates the drill axis from intended position. Such deviation causes drill wandering, significant burrs, and irregular hole diameters. The hole quality is improved significantly when drilling on a polished surface at the same or more aggressive drilling parameters. An improvement of 27% in hole center deviation is achieved for the polished surface of CP titanium (**Figure 29**). Similarly, an improvement of 260% on hole deviation is achieved when drilling a polished PEEK block.

Vickers microhardness near a drilled surface is obtained to study the level of plastic deformation and work hardening below a drilled surface. The hardness near the drilled surface is found to be 15% higher than those at the unmachined zone (**Figure 30**). The work hardening effect is caused by plastic deformation of the

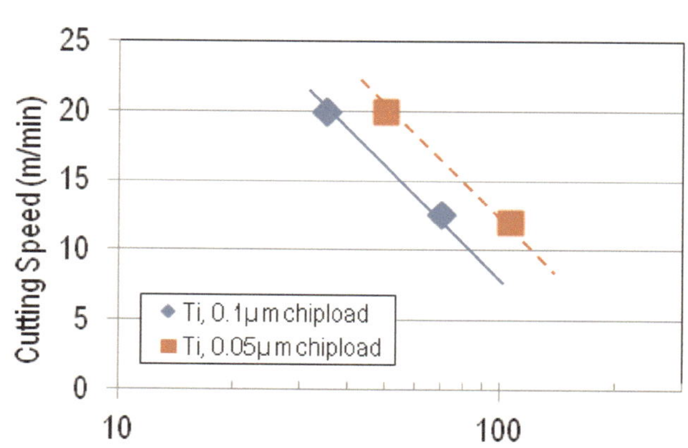

Figure 27.
Tool life plot for microdrilling of CP titanium. Progressive pecking, tool life criterion 8 μm [28].

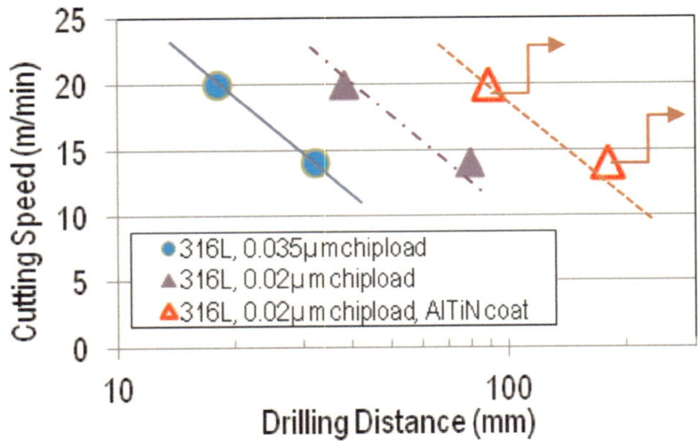

Figure 28.
Tool life plot for microdrilling of 316L stainless steel. Progressive pecking, tool life criterion 15 μm. Drilling with AlTiN coated drills were stopped due to shortage of materials [28].

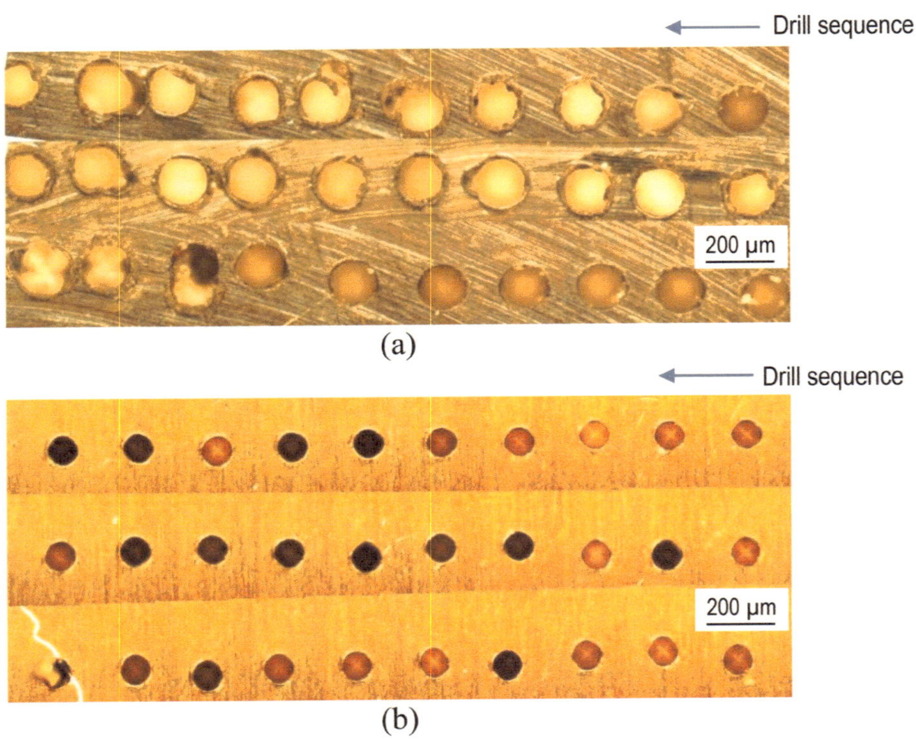

Figure 29.
Composite optical images showing drill wandering and hole accuracy on (a) milled CP titanium surface; 4 m/min, 1 μm/flute, 2:1 aspect ratio, and (b) polished CP titanium surface; 12 m/min, 0.05 μm/flute, 10:1 aspect ratio [28].

surface by a worn tool, and smearing of BUE on the drill wall. Similar work hardening effect is reported while drilling austenitic stainless steel leading to a higher resistance near the chisel edge of the drill. Ideally, the drill cutting edges should not shear the workpiece within the work hardened layer from a previous cut. Since the work hardening zone is about 30 μm, it is impractical to microdrill 316L stainless steel at an aggressive chip load more than 30 μm/flute since a fragile microdrill would simply fracture.

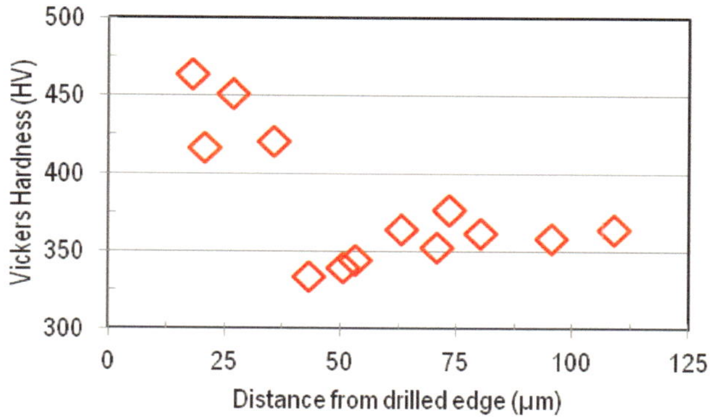

Figure 30.
Vickers microhardness below drilled surface of the 10th hole. 12.7 mm drilling distance, 14 m/min, 0.035 µm/ flute chip load on 316L stainless steel. Microhardness test at 50 g load, 14 s dwell time [28].

4.3 Ultraprecision turning

Product miniaturization and demand for ultraprecision products drives the rapid development of micro/nano scale turning or ultraprecision turning. This technology produces polished and high quality spherical, aspherical parts from metals, ceramics, semiconductors, and polymers that cannot be economically produced by traditional grinding, lapping, or polishing processes. Micro/nano turning also produces intricate shape with low or no subsurface damage since it operates in the ductile-regime mode.

Commercial lathe systems for ultraprecision machining are available. Although tool and axes motions can be in the nanometer ranges, it is an engineering challenge to control the thermal drift issue of a large system. Having a very compact 200-mm system, however, would compromise the required resolution for precision microturning [29]. Diamonds are commonly used for micro/nano turning. Polycrystalline diamond tools are sintered from micron-sized diamond grains. It is less expensive but with limited capability due to large edge radius (few hundred nanometers) and lower edge strength due to attrition wear. Single crystalline diamond tools are best for micro/nano turning since they:

- Have single crystalline structure that allows a sharp cutting edge as small as few nanometers (**Figure 1b**),
- Have highest thermal conductivity among all engineering materials,
- Retain high strength and hardness at high temperature,
- Possess high elastic and shear moduli to resist plastic deformation, and
- Exhibit a low coefficient of friction.

A diamond tool, however, is costly and brittle. A tool with zero or negative rake angle (i) improves its edge strength, and (ii) forms a hydrostatic compressive stress field in the material just in front and below a tool, therefore, minimizes crack initiation. The single crystal diamond typically has (110) crystal plane as rake face and is brazed onto a steel shank of different shape and size.

Not any material can be successfully micro/nano turned with a diamond tool. Ferrous alloys and silicon carbide (SiC) are not suitable for diamond turning

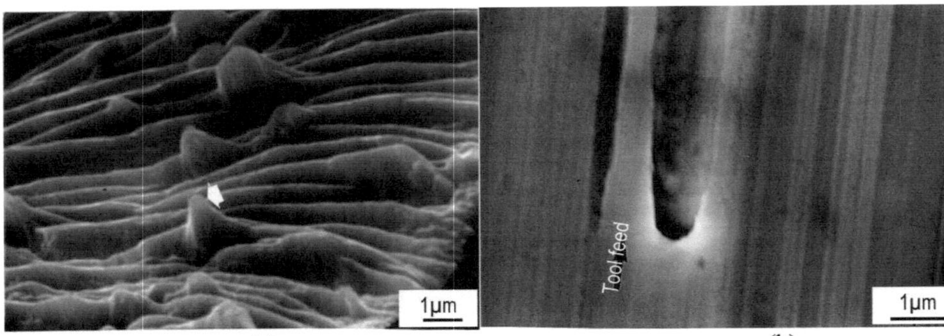

(a) (b)

Figure 31.
(a) Microchip from CA 173 showing a beryllide inclusion, (b) deep scratch on machined surface by a broken beryllide.

because of diffusion from highly concentrated carbon in diamond tool to a lower concentration zone of carbon in workpiece materials when the cutting zone is at high temperature during machining. Selected materials that can be successfully machined with a diamond tool are shown in **Table 4**. These material should be homogeneous and contain few if no impurities. The hard inclusions might either damage a sharp diamond edge or being sheared off and smearing against the machined surface. **Figure 31** shows the hard beryllides in beryllium copper CA173 that plow and smear the mirror finish surface.

Semiconductor	Metal	Ceramic	Plastics
Cadmium telluride	Aluminum alloys	Aluminum oxide	Acrylic
Gallium arsenide	Copper alloys	Zirconium oxide	Fluoroplastics
Germanium	Electroless nickel	Optical glasses	Nylon
Lithium niobate	Gold	Quartz	Polycarbonate
Silicon	Magnesium		Polymethylmethacrylate
Silicon nitride	Silver		Propylene
Zinc selenide	Zinc		Styrene
Zinc sulphide			

Table 4.
Examples of diamond machinable materials.

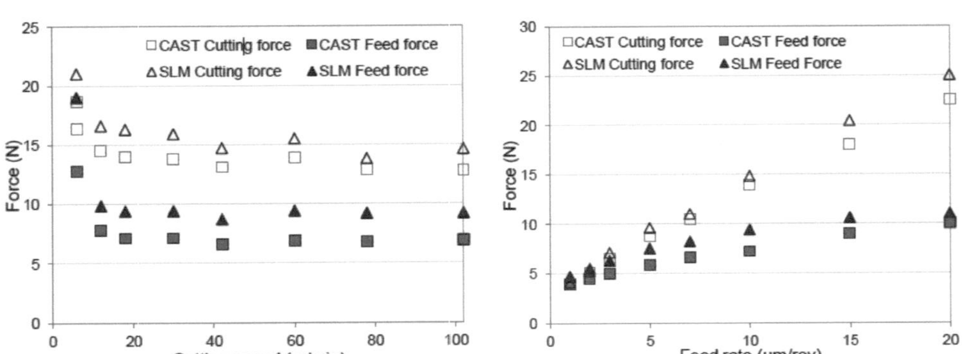

(a) (b)

Figure 32.
Comparison of cutting forces in orthogonal microturning of cast and selective laser melted Ti 6Al 4V [30].

Micromist is required to lubricate and cool both tool and machined surface. A micromist nozzle should move with a tool while blowing micro/nano chips away from the machined surface.

An experimental study was done to compare machinability of Ti 6Al 4V produced by casting or selective laser melting. TiAlN coated microturning tools with 8° rake angle, 1.3 μm edge radius were used. Microturning at orthogonal condition was performed at 6–600 m/min, 1–20 μm/rev, and 500 μm depth of cut. The chip morphologies are similar for both materials and there is no significant effect on microstructure; however, the cutting and feeding forces on AM alloy is about 3–24% higher than those for cast alloy when varying the cutting speed (**Figure 32a**) or changing the feed (**Figure 32b**). Such higher forces would shorten the tool life and subsequently degrade the surface quality.

5. Summary

The demanding for product miniaturization and increasing part precision has fueled the development of micromachining. Recent explosion of product innovation with printing 3D metal parts also escalate the post processing studies due to inherent defects of 3D printed metals. The synergy of subtractive processes—micromilling, microdrilling, and microturning—with additive processes for metals—power bed fusion, material jetting, binder jetting, and direct energy deposition—will enable the successful manufacturing of complex metal parts to meet strict engineering requirement.

Author details

Wayne N.P. Hung[1]* and Mike Corliss[2]

1 Texas A&M University, College Station, Texas, USA

2 KGSBO, Katy, Texas, USA

*Address all correspondence to: hung@tamu.edu

References

[1] Chu WS, Kim CS, Lee HT, Choi JO, Park JI, Song JH, et al. Hybrid manufacturing in micro/nano scale: A review. International Journal of Precision Engineering and Manufacturing-Green Technology. 2014;**1**(1):75-92

[2] Kaynak Y, Kitay O. The effect of post-processing operations on surface characteristics of 316L stainless steel produced by selective laser melting. Additive Manufacturing. 2019;**26**:84-93

[3] Kaynak Y, Tascioglu E. Finish machining-induced surface roughness, microhardness and XRD analysis of selective laser melted Inconel 718 alloy. Procedia CIRP. 2018;**71**:500-504

[4] Jagadesh T, Samual GL. Investigation into cutting forces and surface roughness in micro turning of titanium alloy using coated carbide tool. Procedia Materials Science. 2014;**5**:2450-2457

[5] Afazov AM, Rachev SM, Segal J. Modelling and simulation of micro-milling cutting forces. Journal of Materials Processing Technology. 2010; **210**:2154-2162

[6] Rahim EA. Tool failure modes and wear mechanism of coated carbide tools when drilling Ti-6Al-4V. International Journal of Precision Technology. 2007;**1** (1):30-39

[7] Zhou L, Ni J, He Q. Study on failure mechanism of the coated carbide tool. International Journal of Refractory Metals and Hard Materials. 2007;**25**:1-5

[8] Rutherford KL, Hutchings IM. A micro-abrasive wear test, with particular application to coated systems. Surface and Coatings Technology. 1996; **79**:231-239

[9] Kumar M, Dotson K, Melkote SN. An experimental technique to detect tool–workpiece contact in micromilling. Journal of Manufacturing Processes. 2010;**12**:99-105

[10] Shunmugavel M, Polishetty A, Nomani J, Goldberg M, Littlefair G. Metallurgical and machinability characteristics of wrought and selective laser melted Ti-6Al-4V. Journal of Metallurgy. 2016;**2016**:1-10. Article ID 7407918. Available from: http://dx.doi.org/10.1155/2016/7407918

[11] Gu J, Barber G, Tung S, Gu RJ. Tool life and wear mechanism of uncoated and coated milling inserts. Wear. 1999; **225–229**:273-284

[12] Wang Z, Kovvuria V, Araujo A, Bacci M, Hung NP, Bukkapatnam STS. Built-up-edge effects on surface deterioration in micromilling processes. Journal of Manufacturing Processes. 2016;**24**:321-327

[13] Klocke F, Maßmann T, Gerschwiler K. Combination of PVD tool coatings and biodegradable lubricants in metal forming and machining. Wear. 2005; **259**:1197-1206

[14] Nam JS, Kim DH, Chung H, Lee SW. Optimization of environmentally benign micro-drilling process with nanofluid minimum quantity lubrication using response surface methodology and genetic algorithm. Journal of Cleaner Production. 2015;**102**: 428-436

[15] Ucun I, Aslantasx K, Bedir F. The effect of minimum quantity lubrication and cryogenic pre-cooling on cutting performance in the micro milling of Inconel 718. Proceedings of the Institution of Mechanical Engineers, Part B: Journal of

Engineering Manufacture. 2015;**229** (12):2134-2143

[16] Hung NP, Fu YQ. Effect of crystalline orientation in the ductile-regime machining of silicon. Journal of Advanced Manufacturing Technology. 2000;**16**:871-876

[17] Kajaria S, Chittipolu S, Adera S, Hung NP. Micromilling in minimum quantity lubrication. Machining Science and Technology. 2012;**16**:524-546

[18] Ziberov M, Bacci da Silva M, Jackson M, Hung NP. Effect of cutting fluid on micromilling of Ti-6Al-4V titanium alloy, NAMRC 44-129. Procedia Manufacturing. 2016;**5**:332-347

[19] Berestovskyi D, Hung NP, Lomeli P. Surface finish of ball-end milled microchannels. Micro- and Nano-Manufacturing. 2014;**2**(0411005):1-10

[20] Rysava Z, Bruschi S. Comparison between EBM and DMLS Ti6Al4V machinability characteristics under dry micro-milling conditions. Materials Science Forum. 2016a;**836-837**:177-184

[21] Kuram E, Ozcelik B. Multi-objective optimization using Taguchi based grey relational analysis for micro-milling of Al 7075 material with ball nose end mill. Measurement. 2013;**46**:1849-1864

[22] Sadiq M, Hoang MN, Valencia N, Obeidat S, Hung NP. Experimental study of micromilling selective laser melted Inconel 718 superalloy. Procedia Manufacturing. 2018;**26**:983-992

[23] Vazquez E, Gomar J, Ciurana J, Rodríguez CA. Analyzing effects of cooling and lubrication conditions in micromilling of Ti6Al4V. Journal of Cleaner Production. 2015;**87**:906-913

[24] Khan WA, Hoang MN, Tai B, Hung NP. Through-tool minimum quantity lubrication and effect on machinability. Journal of Manufacturing Processes. 2018;**34**(Part B):750-757

[25] Zhang Y, Jun MBG. Feasibility of lignin as additive in metalworking fluids for micro-milling. Journal of Manufacturing Processes. 2014;**16**: 503-510

[26] Bruschi S, Tristo G, Rysava Z, Bariani PF, Umbrello D, De Chiffre L. Environmentally clean micromilling of electron beam melted Ti6Al4V. Journal of Cleaner Production. 2016;**133**:932-941

[27] Rysava Z, Bruschi S, Carmignato S, Medeossi F, Savio E, Zanini F. Micro-drilling and threading of the Ti6Al4V titanium alloy produced through additive manufacturing. Procedia CIRP. 2016b;**46**:583-586

[28] Mohanty S, Wells S, Hung NP. Microdrilling of Biocompatible Materials. IMECE2012-87523, Proceedings, ASME International Mechanical Engineering Congress & Exposition; Houston, Texas. 2012

[29] Lu Z, Yoneyama T. Micro cutting in the micro lathe turning system. International Journal of Machine Tools and Manufacture. 1999;**39**:1171-1183

[30] Coz GL, Fischer M, Piquard R, D'Acunto A, Laheurte P, Dudzinski D. Micro cutting of Ti-6Al-4V parts produced by SLM process. Procedia CIRP. 2017;**58**:228-232

Chapter 2

Pico- and Femtosecond Laser Micromachining for Surface Texturing

Tatsuhiko Aizawa and Tadahiko Inohara

Abstract

The pico- and femtosecond laser micromachining has grown up as a reliable tool for precise manufacturing and electronic industries to make fine drilling and machining into hard metals and ceramics as well as soft plastic and to form various nano- and microtextures for improvement of surface functions and properties in products. The ultrashort-pulse laser machining systems were developed to describe the fine microdrilling and microtexturing behavior for various materials. Accuracy in circularity and drilled depth were evaluated to discuss the effect of substrate materials on the laser microdrilling. Accuracy in unit geometry and alignment was also discussed for applications. A carbon base mold substrate was micromachined to transcribe its microtextures to transparent plastics and oxide glasses. Three practical examples were introduced to demonstrate the effectiveness of nano-/microtexturing on the improvement of microjoinability, the reduction in friction and wear of mechanical parts and tools, and the surface property control. The fast-rate laser machinability, the spatial resolution in laser microtexturing as well as the laser micromanufacturing capacity were discussed to aim at the future innovations in manufacturing toward the sustainable society.

Keywords: picosecond laser micromachining, femtosecond laser micromachining, microdrilling, microtexturing, nano-/microtexturing, laser micropatterning

1. Introduction

The laser technology for manufacturing is classified into two categories; e.g., thermal and athermal processings [1]. CO_2-laser with continuous power supply and fiber-lasers with use of short pulses are typical processing for welding, machining, and joining by formation of thermally hot spots [2]. Various fiber lasers [3] have been developed and applied to laser welding, laser machining, laser marking, and so on [3]. Most of them utilize the nanosecond solid-state oscillators and make thermal machining of materials. In recent, pico- and femtosecond laser machining [4–6] is widely utilized for athermal removal of materials with high dimensional accuracy in practice.

There are two keywords to classify the laser processing; i.e., the wave length of light and the pulse duration time. CO_2 laser has the longest wave length of 10.6 μm, while excimer laser by KrF, 248 nm. Most of laser wave length (λ) ranges from the far ultra-violet regime less than 200 nm to infra-red regime more than 20 μm. Since every material has its own relaxation rime (τ_0), most of laser power can be absorbed by the material having the equivalent τ_0 to λ. Then, this targeting work material is athermally

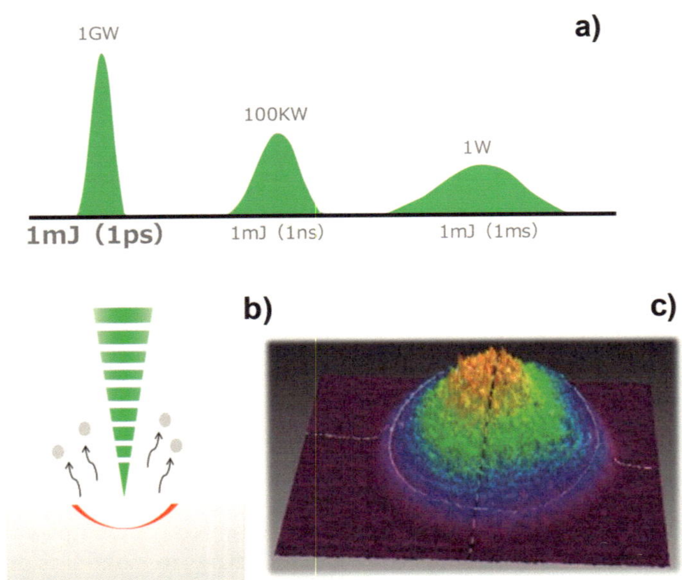

Figure 1.
Typical characteristics of ultra-short pulse laser machining. (a) Significant increase of laser power by reduction of Δt down to 1 ps, (b) ablation as an athermal removal of materials, and (c) laser intensity profile.

machined by selecting the laser with suitable wave length; otherwise, the work is only thermally cut or drilled. A micromachining essentially requires for fast-rate removal of materials with sufficient accuracy in dimension and geometry; the repetition frequency as well as the wave length must be optimally selected to make suitable laser micromachining to each work-material. With use of second harmonic generator (SHG), third harmonic generator (THG), and forth harmonic generator (FHG), the fundamental wavelength of 1064 nm is controllable to be 532, 355 and 266 nm, respectively.

The pulse duration time (Δt) is important for short-pulse laser micromachining. As shown in **Figure 1a**, the pulse power increases significantly with reduction of Δt. When the laser energy with Δt = 1 ms is 1 mJ, the laser power (P) is only 1 W; P reaches to 1 GW only by shortening Δt down to 1 ps.

Under high power laser irradiation, most of materials are athermally removed, or, ablazed, as depicted in **Figure 1b**. The dimensional accuracy in laser micromachining is determined by focusing the laser spot for this ablation process. This laser irradiation has a finite spot size which is dependent on λ and Δt. The laser intensity distributes even in the focused spot; e.g., the well-controlled laser intensity distributes in Gaussian profile as depicted in **Figure 1c**.

In the following, our developing ultrashort pulse laser machining systems are employed to make microdrilling and microtexturing into various kinds of work materials. In particular, the laser microtexturing technology is applied to micro-joining process of dissimilar polymers, and to microdimple formation for friction control of sliding parts and components and for reservoir of wear debris during dry cutting. Further applications including the surface property control by using the nano-/microtexturing are discussed in this chapter.

2. Pico- and femtosecond laser micromachining system

Our developed pico- and femtosecond laser machining systems are stated with some comments on their capacity and configuration.

2.1 Picosecond laser micromachining system

A single picosecond is equivalent to the relaxation time of molecular bonding stage; its pulsed power is readily absorbed by most of work materials. Three types of picosecond laser machining systems were developed; a standard system and its configuration are shown in **Figure 2**. The machining speed is dependent on the repletion frequency and average power. The dimensional accuracy in machining is determined by the beam spot size. To be discussed later, this spot size depends on the optical system; e.g., the minimum spot size can be controlled down to 1 μm when every lens is fixed on the stage. However, when the lens position is controlled during machining, the spot size becomes wider; e.g., it is limited by 10 μm when using the galvanometer.

2.2 Femtosecond laser micromachining system

A single femtosecond or subpicosecond lasers are developed for innovative research and development; most industrial applications stand on this laser machining in the order of 100 femtoseconds. Our developed system and its configuration are shown in **Figure 3**.

Since the focused spot of work materials is subjected to ultra-high power irradiation, how to scan the beam spot becomes more important when using this laser machining system. Higher repetition frequency of laser beams as well as higher scanning speed result in fast-rate dimensionally accurate machining. At present, a laser oscillator with the repetition frequency of 40 MHz has been already developed for machining. How to make fast control of this short pulse laser becomes an essential issue in laser machining design.

Wave length	515nm
Pulse duration	<10ps
Average power	30W
Repetition	MAX200kHz
Beam spot size	MINΦ10μm using Galvanometer MINΦ1μm using fixed lens
Stage area	400mm×400mm
Sample Weight	< 10kg

Figure 2.
Our developed picosecond laser machining system and its capacity and configuration.

Wave length	515nm
Pulse duration	180fs～190fs
Average power	8.2W
Repetition	MAX600kHz
Beam spot size	MIN:10μm using Galvanometer MIN:1μm using fixed lens
Stage area	300mm×300mm
Scanning speed	MAX 3000mm/s
Scanning area	MAX 30mm×30mm
Sample weight	< 30kg

Figure 3.
Our developed femtosecond laser machining system and its capacity and configuration.

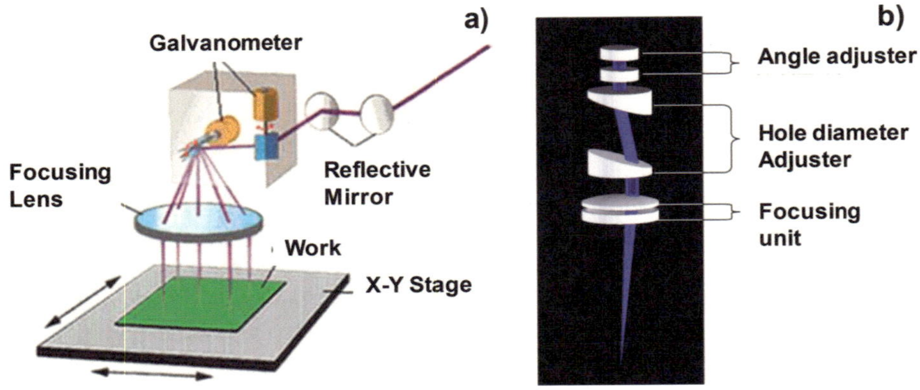

Figure 4.
Typical two optical control units for laser micromachining. (a) Optical control unit with use of galvanometer and (b) beam rotator.

2.3 Optical unit control

In parallel with the development of laser oscillators and machining unit, the optical unit design is also important for accurate laser machining. Two unit designs are introduced in **Figure 4**; e.g., an optical control unit with use of galvanometer to distribute the laser beam as designed, and a beam rotator for laser drilling with accurate circularity. The former unit is a standard approach for laser machining with moderate rate; new controller must be developed to make much faster rate laser machining. The latter is a powerful tool to rotate the optical units and to move the laser beam in the axisymmetric manner.

Various controlling tools of laser beam can be designed and developed for each application of laser machining.

3. Fine microdrilling into metallic alloys and ceramics

These pico- and femtoseconds with the pulse duration in the order of 10^{-12} and 10^{-15} s provide a reliable means to drill the through-holes into the ceramics, the metallic alloys, and the plastics [7]. Compared to the micromilling and the microelectrical discharge machining (micro-EDM), finer through-holes with higher circularity are formed without residuals at the inlet of holes and without deterioration on their inner surfaces [8]. In addition, no micromilling tools and no thin EDM wires are needed to drill the lots of through-holes onto the relatively large area. In this laser drilling process, the surface quality of through-holes as well as their circularity is strongly dependent on the laser beam control, as summarized in [9]. In the conventional fiber-laser machining, the inlet of through-holes is deteriorated by the redeposits and the residuals [10]. Even when using the picosecond pulse lasers, the through-hole shape is also damaged by the unstable laser beams [11, 12]. Typical damage of through-holes comes from the branching from the straight hole drilled in the initial stage to two holes. The deviation of beam focusing and positioning directly induces these defects [7, 8, 13]. Our developed picosecond laser machining system for industrial applications is applied to drill the through-holes into the ceramic plates. The beam rotator is used as a trepanning system for laser drilling. The alumina plate with the thickness of 1 mm is employed as a substrate. Scanning electron microscopy (SEM) is used to measure the diameter of through-holes as well as their aspect ratio. The replica method is also utilized to describe the geometric alignment and homogeneity of through-holes. The drilled through-holes with the uniform diameter of 50 μm and the aspect ratio of 10.0 are accurately aligned into the alumina plate. The present trepanning device works to control the diameter of irradiation for fine drilling of the tapered and inversely tapered through-holes into steels.

3.1 Microdrilling into ceramic substrates

The alumina plate is prepared for the laser drilling under the experimental setup in **Figure 5**. The beam rotator as well as projection lens unit is utilized to improve the focused beam quality. Through the CCD and display, the microdrilling process is monitored during operation.

Let us first evaluate on the difference of drilling behavior between the fiber lasers and the picosecond laser. The through-hole with the diameter of 50 μm is drilled into alumina plate. When using the normal fiber lasers, the surroundings of hole are completely damaged with deposits on them **Figure 6a**. While, the accurate hole with circularity of 1 μm is drilled by the picosecond laser without damage and deposits, as shown in **Figure 6b**.

No residuals or no redeposits at the vicinity of through-hole inlets prove that the present picosecond lase drilling is free from the thermal effects to deteriorate the surface quality of work specimen. The picosecond laser drilling is utilized to

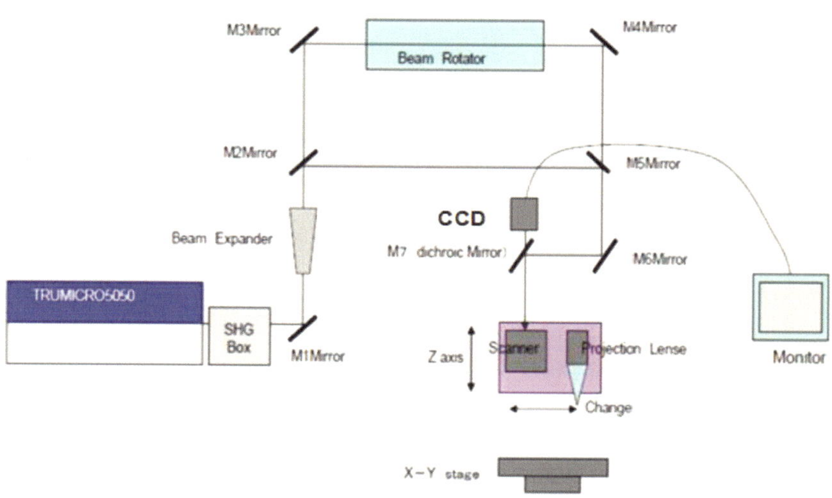

Figure 5.
A typical experimental setup for laser microdrilling.

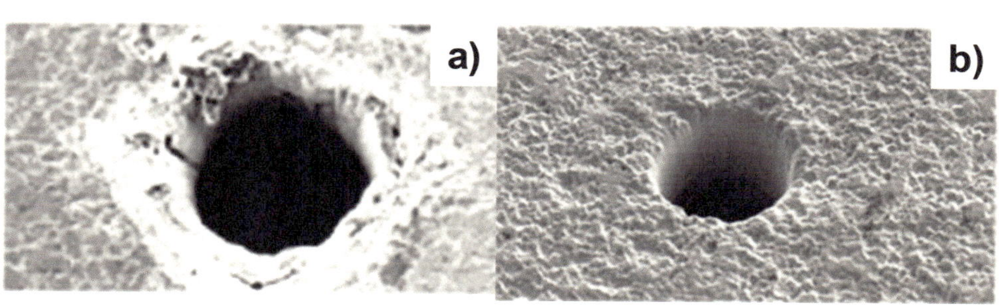

Figure 6.
Comparison of the drilled through-hole between fiber lasers and the picosecond laser. (a) Fiber laser drilling, and, (b) Picosecond laser drilling.

fabricate a series of holes with periodically aligned into alumina substrate. **Figure 7a** depicts the through-holes drilled into the alumina. Each through-hole is aligned with the pitch of 300 μm as programmed by the CAM data mining through the positioning control of beams. As had been discussed in [11], the inner surface quality of through-holes is sensitive to the instability during the laser drilling. **Figure 7b** also demonstrates that the straight through-hole inner surfaces are formed to have constant diameter without any geometric damages by the picosecond laser drilling. This is because the laser beam is well profiled through the trepanning system before fine control by the galvanometer, and is controlled to move into the depth of work materials. The above straightforwardness of through-holes is also demonstrated by using the replica method. In this method, the silicone-based polymers are infiltrated into each through-hole. The frozen polymers are used as a replica to reproduce the drilled through-hole shape. **Figure 7c** depicts the alignment of replicas in correspondence to a series of laser-drilled through-holes. Three through-holes were laser-drilled down to the same depth in the alumina plate. Since the first three polymer pillars have the same height as 150 μm, the successive series of through-holes are accurately machined into the alumina with the same depth.

These straight through-holes with high aspect ratio provide a solution to the demand for the probe-cards to make accurate inspection of the semiconductor

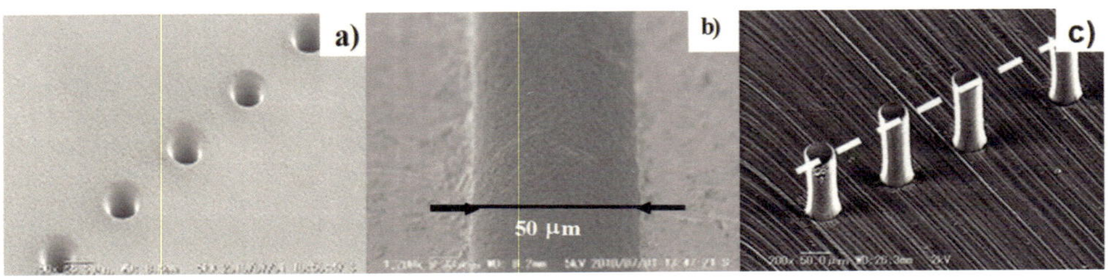

Figure 7.
Picosecond laser drilling of through-holes into the alumina plate. (a) Alignment of through-holes, (b) inner surface of through-hole with the diameter of 50 μm, and (c) demonstration of the homogeneous laser drilling by using the replica method.

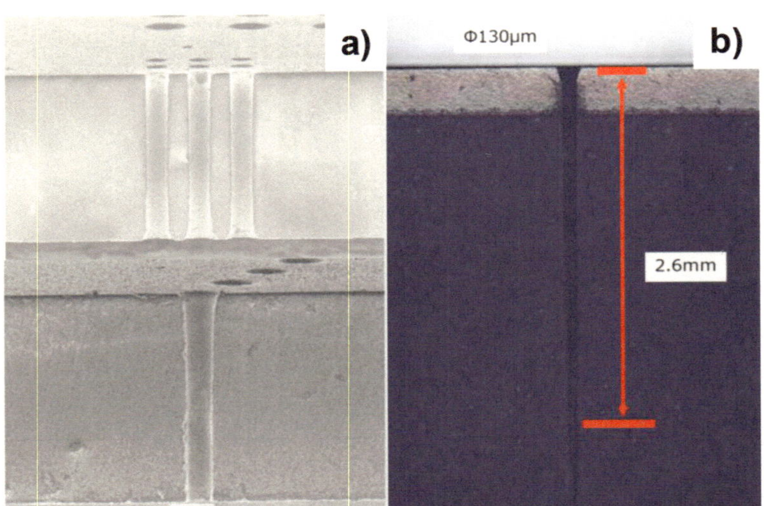

Figure 8.
Picosecond laser drilling of through-holes with higher aspect ratio. (a) Drilled through-holes into alumina plate and (b) drilled through-holes into partially stabilized zirconia (PSZ).

chips. The probe-pins are pierced through the straight through-holes of alumina or PSZ substrates for inspection. These through-holes must have higher aspect ratio than 10 to preserve the sufficient working space. **Figure 8a** depicts the through-hole with the diameter of 50 μm machined into the alumina plate with the thickness of 1 mm; the aspect ratio reaches to 20. This high aspect ratio is also attained even when laser drilling PSZ in **Figure 8b**. This demonstrates that the trepanned laser drilling enables to make through-holes with higher aspect ratio than 20 under the well-structured setup in laser machining.

3.2 Microdrilling into metallic alloys

In the die and mold industries, the case-hardened and plasma-treated steels are often utilized for high proof of dimensional accuracy. Let us also compare the laser drilling performance between the fiber-lasers and the picosecond laser. **Figure 9** compared the drilled through-holes between two lasers. The large heat-affected zones as well as damages surround the drilled hole by fiber laser in **Figure 9a**. While, the clean and accurate through-hole is drilled into the case-hardened steels by the picosecond laser in **Figure 9b**.

Without use of the beam rotation control, the tapering is difficult or nearly impossible in the laser drilling. In the present setup, the pair of lenses in the beam rotator in **Figure 5** is radially adjusted to directly control the diameter of irradiated

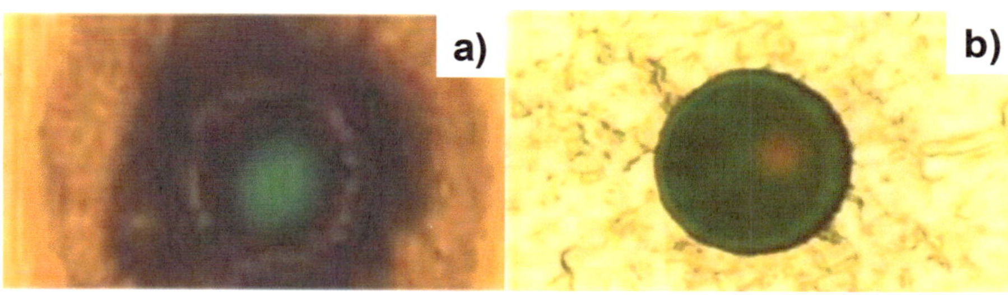

Figure 9.
Comparison of the drilled through-holes into the case-hardened steels. (a) Using the fiber lasers and (b) using the picosecond laser.

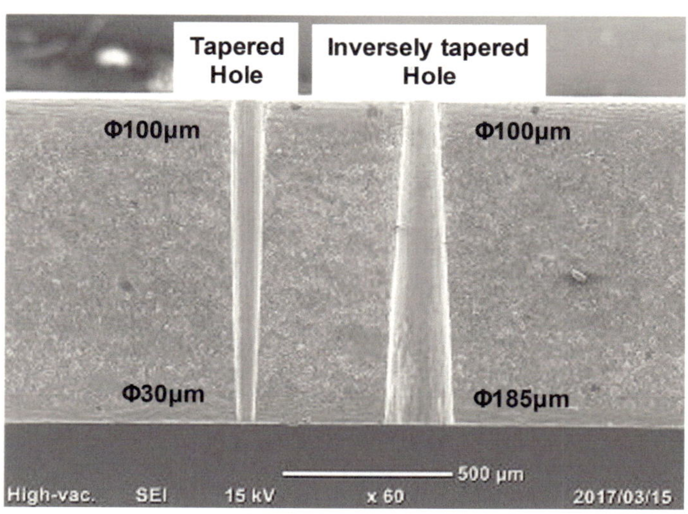

Figure 10.
Picosecond laser drilling of the tapered and inversely tapered through-holes into the case-hardened steels.

region. When this diameter is narrowed from the inlet to the outlet with the constant velocity, the uniformly tapered through-hole is drilled to have a constant tapered angle up to the specified positive skew angle. On the other hand, the inversely tapered through-hole is also machined by enlarging this diameter with the constant velocity in the similar way down to the negative skew angle. These tapering or inversely tapering processes from the inlet to outlet of the through-hole are automatically programmed. After CAM data in the present laser drilling, this diameter of irradiation is narrowed from the inlet by 100 μm to the outlet by 30 μm with the constant feeding velocity. Then, the tapered through-hole is built into the alumina plate with the constant angle of +30° and the higher aspect ratio than 10.0 in **Figure 10**. In the similar way, the inversely tapered through-hole is formed by enlarging the diameter of irradiation from the inlet by 100 μm to the outlet by 180 μm also with the constant velocity. The inversely tapered through-hole is also drilled into the alumina with the thickness of 1 mm. The inversely tapered angles are also constant by −25°. In both cases, the inner surfaces of holes are finely shaped with less roughness [14].

4. Fine microtexturing onto metallic alloys and polymers

Microtextures with the size in the order of 1–100 μm on the solid surface and interface work to reduce the friction and wear, to assist the joinability, and to functionalize the surfaces and interfaces [15]. Micromilling [16] and microelectrical discharge machining (micro-EDM) [17] have been utilized to make microtexturing onto the steel surfaces. Due to the limitation on the machining tool shape and their controllability for machining, their application is also limited in practice. Short-pulse laser machining is employed to make microtexturing onto the metallic and ceramic surfaces.

4.1 Microtexturing of dimples onto the surfaces

A circular dimple is formed on the various metallic surfaces as an aligned structure. **Figure 11** depicts four microtexturing cases. The unit-geometry of microdimples, their alignment on the surfaces, and the finished surface quality are preserved with less roughing during laser processing. For an example, the circular microdimples with the diameter of 95 μm and the depth of 26 μm are formed on the AISI430 surface in the pitch of 110 μm as shown in **Figure 11d**. No difference

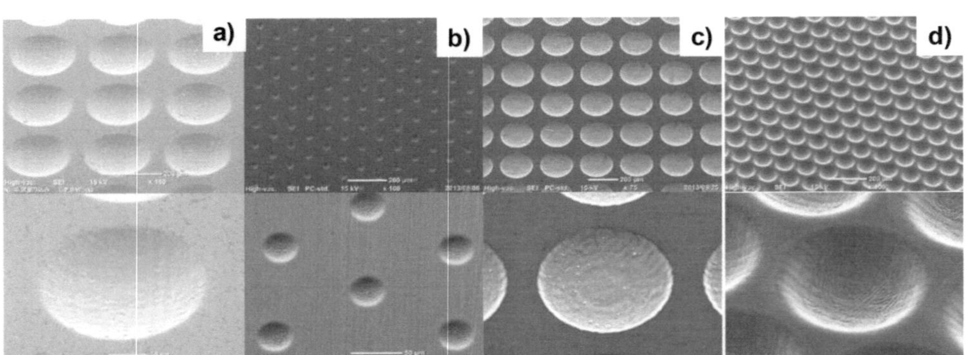

Figure 11.
Laser microtexturing of circular dimples aligned on the metallic substrates. (a) Aluminum, (b) copper, (c) nickel, and (d) AISI430.

in microdimple size and shape and in its alignment is noticed for various kinds of metallic substrates.

4.2 Microtexturing of embosses onto the surfaces

The initial geometric data in CAD and CAM for laser microdimple texturing are data-transformed from positive to negative; this transformed CAD and CAM data are automatically built for laser microemboss formation. In practice, the concave patterning to form the microdimples changes itself to the convex patterning to form the microembosses onto the substrate surfaces. **Figure 12** depicts four microembossing cases. The dimensional and geometric accuracies are preserved in the similar manner of microdimple formation. For an example, the circular microembosses with the diameter of 250 μm and the depth of 125 μm are formed on the boron-silicate glass surface in the pitch of 280 μm, as shown in **Figure 12d**.

4.3 Microtexturing of 3D-lattice structures onto the surfaces

With use of femtosecond lasers, finer microtextures are formed as a three-dimensional structure on the metallic surfaces. **Figure 13** depicts the three-dimensional microstructures formed on the steel surfaces. In particular, the

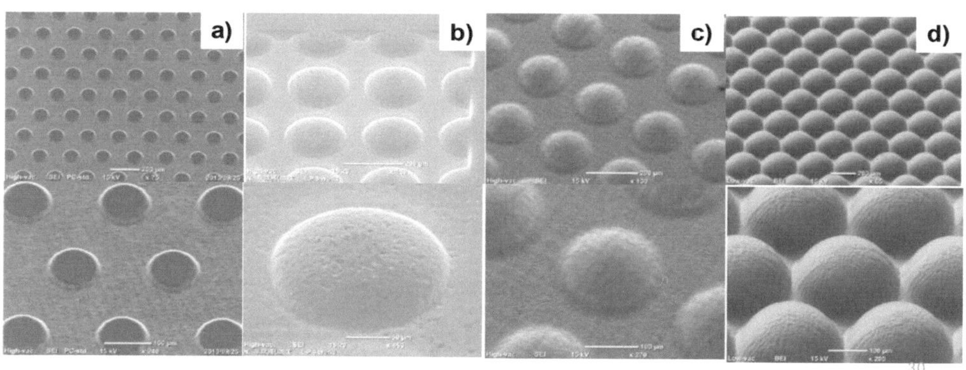

Figure 12.
Laser microembossing of circular embosses aligned on the metallic and ceramic substrates. (a) AISI430, (b) Ni, (c) AISI304, and (d) boron-silicate glass.

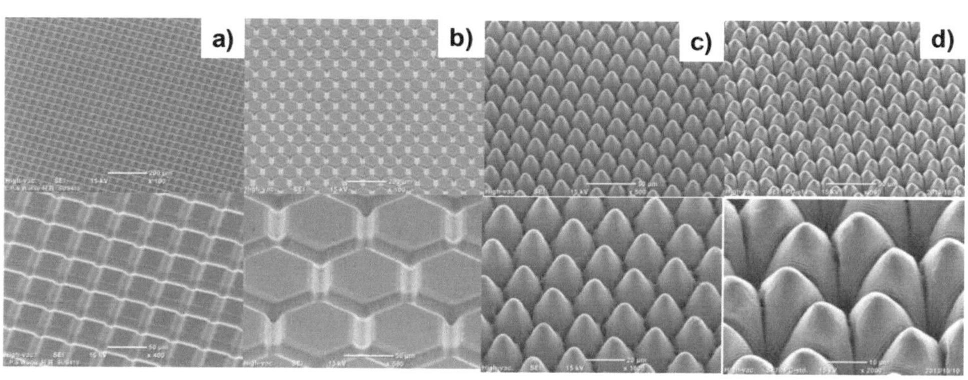

Figure 13.
Laser microtexturing of three-dimensional structures onto the surfaces. (a) AISI410, (b) SISI304, (c) AISI430, and (d) AISI430.

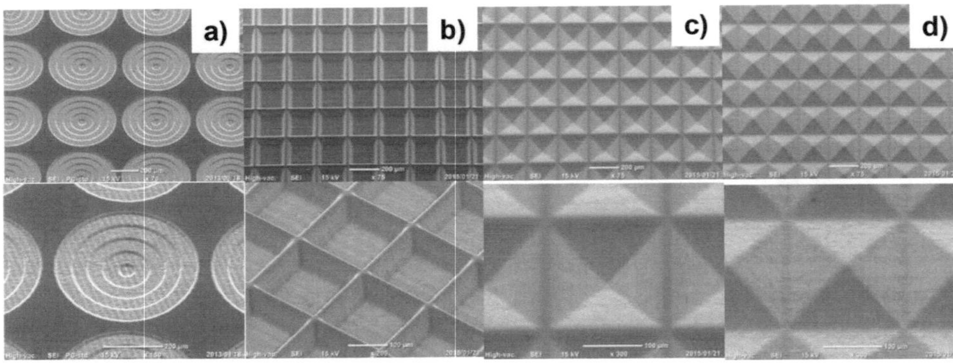

Figure 14.
Laser microtexturing of fine periodic structures onto the surfaces. (a) Al, (b) AISI304, (c) AISI304, and (d) AISI304.

Gaussian-shaped pillar array with the height of 20 μm and the pitch of 20 μm is machined into the AISI430 substrate as shown in **Figure 13c**.

4.4 Microtexturing of periodic structures onto the surfaces

Three-dimensional periodic microstructures have a capability to functionalize the metallic surfaces for optical reflection and diffraction devices and for stamping die and injection mold to transcribe their negative textures onto metallic and polymer sheets. **Figure 14** depicts the periodic microstructures formed on the aluminum and AISI304 steel substrates, respectively. **Figure 14a** is a stepwise terrace structure machined into aluminum with each layer thickness of 5 μm by decreasing the diameter from 450 μm down to 50 μm with the step of 100 μm.

5. Fine microtexturing into carbon-base molds

Two- and three-dimensional microtexturing becomes much important in preparation of mold-dies for mold-stamping of optical elements [18]. The most popular microtexture is a Fresnel pattern for optical lens; circumferential patterns with steep surfaces must be imprinted onto the surface of substrate materials. V-letter-shaped micropatterns are laser-machined onto the glassy carbon substrate to discuss the dimensional accuracy and to investigate the depth profile for different aspect ratio. Furthermore, our developing microstamping system [8, 19, 20] is utilized to duplicate these micropatterns onto optical polymers by using the patterned glassy carbon mold-dies and to discuss the accuracy by this imprinting.

5.1 Microtexturing into glassy carbon die substrate

In the two-dimensional microtexturing, a unit pattern like a groove, a dimple, or a wedge is machined with the specified regularity onto the substrate by using X-Y positioning control. Here, a microgroove is employed as a standard unit pattern to fabricate the microtextured mold-die. Glassy carbon (GC) substrate is employed to make V-letter-shaped microgrooving with the pitch of 35 μm, the V-shaped wedge width of 10 μm, and the depth of 10 μm in design. **Figure 15a** shows the optical micrograph of V-shaped grooving pattern on GC substrate. One groove is laser-machined twice on the same designed machining path. This micropattern is formed onto the GC substrate with the area of 25 × 25 mm^2 for 40 min or 2.4 ks. As shown in **Figure 15b**, a sharp wedge of microgroove is imprinted onto the multilayered GC substrate. The microgroove has 10 μm in width, and 35 μm in pitch. The geometric dimensions specified in CAM program are accurately reproduced in the actual laser microtexturing. The depth profile of V-letter-shaped microgrooves is directly measured to investigate the accuracy of depth in the two-dimensional texturing. **Figure 16** depicts the measured depth profile by precise surface profilometer. Deviation of depth ranges from −1 to +2 μm around the average depth of 10 μm. This proves that regular patterns could be machined by the present approach. In order to investigate the controllability of microgrooving in depth, the designed depth parameter is varied with the laser beam power kept constant. **Figure 17** compares the relationship between the designed and

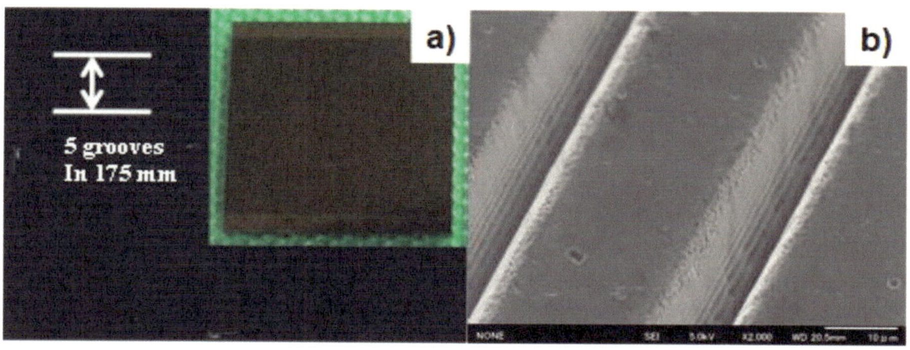

Figure 15.
Laser microtexturing of V-letter shaped grooves into GC substrate. (a) Microscopic image of microgrooved GC and (b) SEM image of microgrooves.

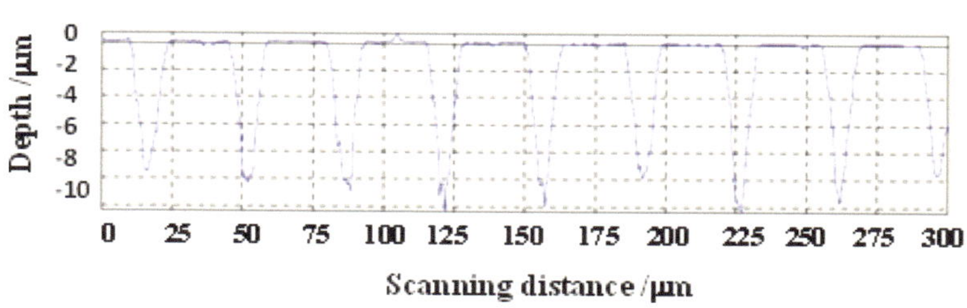

Figure 16.
Depth profile across the V-letter shaped microgrooves in GC substrate.

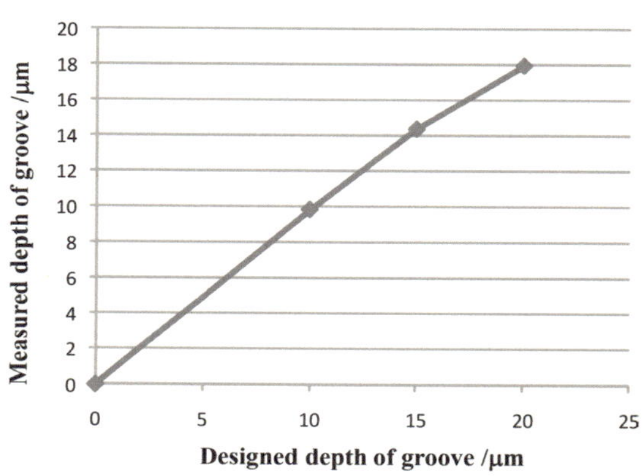

Figure 17.
Relationship between the designed and measured microgroove depths.

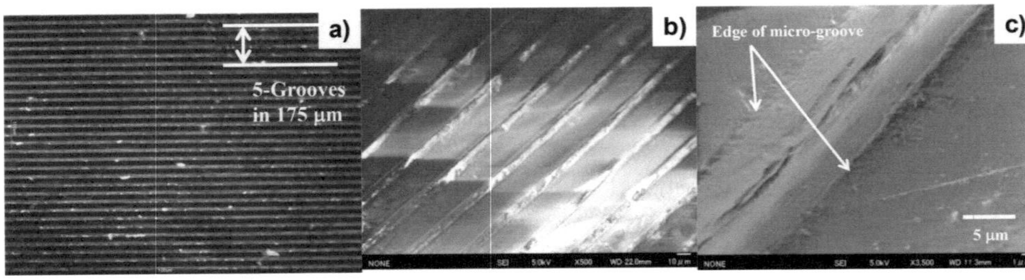

Figure 18.
Transcription of the V-letter wedge microtexture on GC to the V-letter bump microtexture via the mold stamping. (a) Multimicrogrooved PMMA sheet, (b) V-letter bump microtextures on PMMA, and (c) formation of microbump by inclusion of melt PMMA into V-letter wedge on GC mold.

measured depths in this microgrooving. Up to 20 μm, the average depth of microgrooves is accurately controlled by the present laser machining system.

5.2 Mold stamping into optical plastics

The above microtextured GC substrate is used as a mold-die for warm mold-stamping. PMMA sheet with the thickness of 1 mm is employed as a work material for this mold-stamping just above its glass-transition temperature of 383 K (or 110°C). **Figure 18a** showed the V-letter-shaped grooving patterns, which are imprinted onto PMMA by the load of 1 kN for 60 s. The V-letter-shaped concave patterns in **Figure 15b** are accurately imprinted onto PMMA as the convex micropattern as shown in **Figure 18b**. That is, a series of microwedge fins are fabricated by this mold-stamping with use of microtextured mold in **Figure 15**. In the mold-stamping, the filling process of work materials into the micropatterns on the mold-die is essential for accurate imprinting. Precise observation with higher magnification in SEM is made to investigate this filling behavior at the initial stage of mold-stamping. **Figure 18c** depicted a convex bump with the width of 10 μm and the height of 3.5 μm. This bump formation is just the initial stage of filling process for viscous PMMA to infiltrate into the V-letter-shaped groove by mold-stamping. In case of mold-stamping just above the glass transition temperature, viscosity of plastic materials is so high as to reduce the filling velocity. This reflects on the slow shearing along the side faces of microgroove.

6. Microjoining of dissimilar polymers by laser microtexturing

Most of mobile cellular phones are not water-proven so as to be diminished in the accident where those were dropped into water. To be free from these damages, there have been done many efforts to install the perfect waterproof into them [21]; e.g., a silicone rubber ring was sandwiched between plastic cover cases to prevent from water penetration through clearance. This fixture might work well just after shipping; it could be useless at the presence of dirt on the interface or through its misalignment by users in daily use of mobile phones. As the first remedy, a liquid silicone rubber with adhesives is fixed onto their polymer case by the liquid injection molding (LIM) process [22]. Since adhesives invoked in the silicone are responsible for joining, delamination might occur in partial after repetitive opening-and-closing operations in daily use of mobile phones. This difficulty requests us to reconsider the joining process between flexible rubber and hard plastic case in the mobile phone.

The microgrooves are formed into the stainless steel mold for injection molding [23, 24]. Silicone rubber is joined with the polycarbonate plate as a specimen for joining strength test. The measured joining strength is constant by 4 N/mm at the presence of fine microgrooves, where the thinnest silicon rubber fractures without interfacial delamination. This joinability is common to the mobile phone model. The waterproof testing demonstrates that this joined interface has sufficient integrity at high pressure state by 15 kPa.

6.1 Microgrooving into the mold for injection molding

The picosecond laser microtexturing with use of the galvanometer is employed to form the microgroove textures onto the AISI martensitic stainless steel mold. **Figure 19** depicts four microgrooved AISI420 molds with varying widths of 100, 75, 45, and 20 μm, respectively. The groove depth is constant by 10 μm. Each microgroove is shaped to have Gaussian profile irrespectively; the beam intensity profile directly reflects on this microgroove geometry. This mold is inserted into the die-set for injection molding. Polycarbonate (PC) is employed as a work material to imprint these microgroove textures onto the work surface. **Figure 20** depicts the transcribed microbump textures onto PC from the microgroove on the AISI420 mold. Both the groove width and pitch are accurately preserved through this injection molding.

6.2 Joining strength test

In the LIM process, adhesive primer is deposited onto the interface before infiltration of silicone melt in the mold-die. Since intermission between two processes is less than 2–3 s, adhesion takes place between silicon and PC-plate under the cooling stage. **Figure 21a** depicts the PC plate specimen with a silicone square ring after joining in the inside of mold-die during LIM process. In the following test, only the joined section in the width of 80 mm is used for tensile adhesive strength testing. A uniaxial

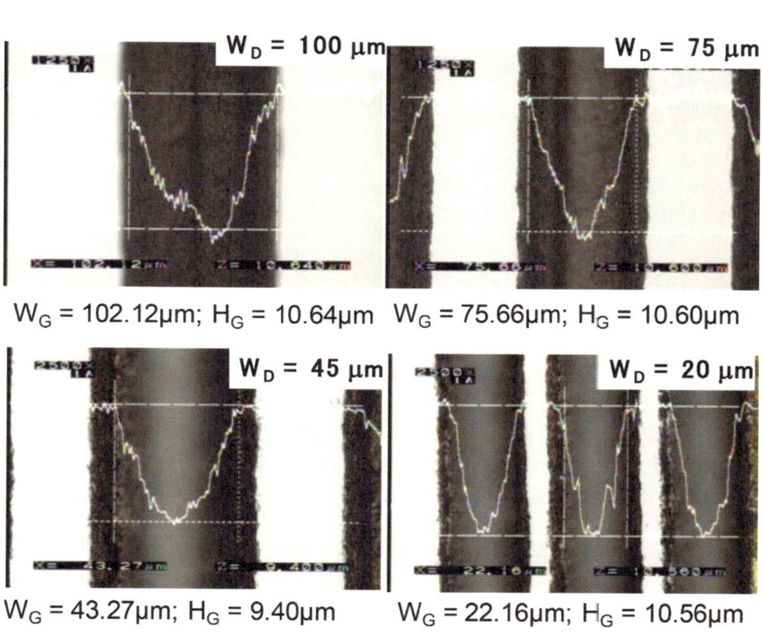

Figure 19.
Microgroove textures with various widths from 100 to 20 μm and constant depth of 10 μm into AISI420 stainless steel substrate.

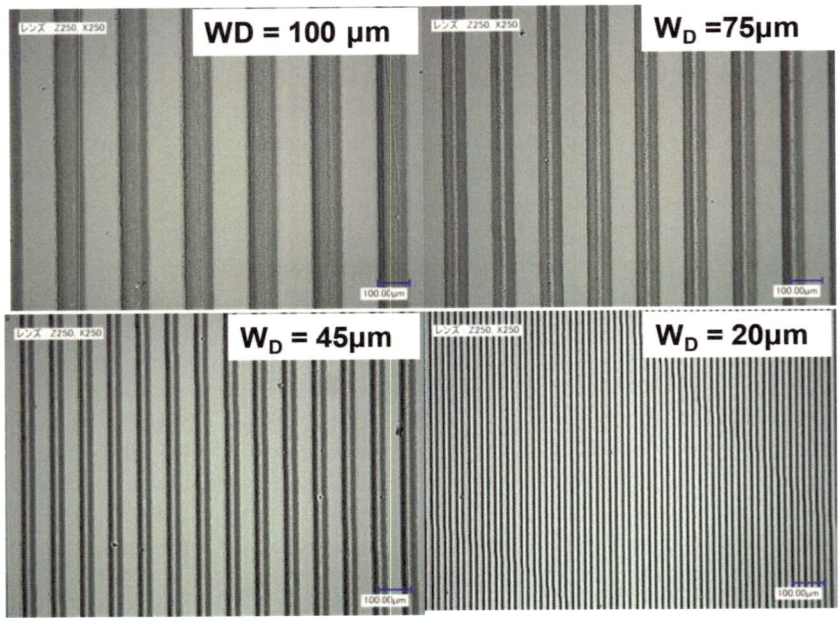

Figure 20.
Transcription from the microgrooves on the AISI420 mold to the microbumps on the PC specimen.

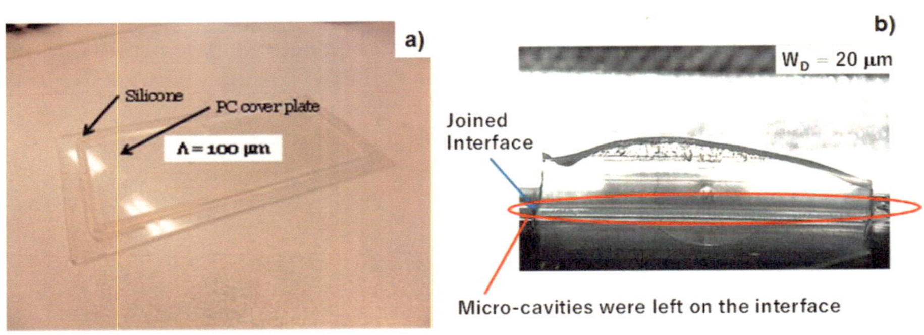

Figure 21.
Joining strength testing. (a) Microbump textured PC specimen joined with silicone rubber and (b) fatal fracture of silicone without interfacial delamination.

tensile testing system with the dynamic video monitoring is used to measure the loading behavior till the final fracture with in situ observation on the deformation of silicone. As shown in **Figure 21b**, when the microgroove width is less than the intrinsic microcavity width of 100 μm, the fatal fracture occurs in the tensile silicone rubber without any delamination of interface between PC and silicone. This joining strength reaches 4 N/mm irrespective of the joined length and size even if the microcavities are present on the interface. This implies that microtextures on the joined interface could control the cavitation process to be free from interfacial delamination.

6.3 Waterproof test of cellular-phone model

The skewed microgrooves with their width and depth of 20 μm are laser-machined into the AISI420 die insert. In the similar way to preparation for the PC-specimen with the silicone rubber ring, the injection molding is used to transcribe the microgrooves into the PC-cover case; LIM process is also utilized to make

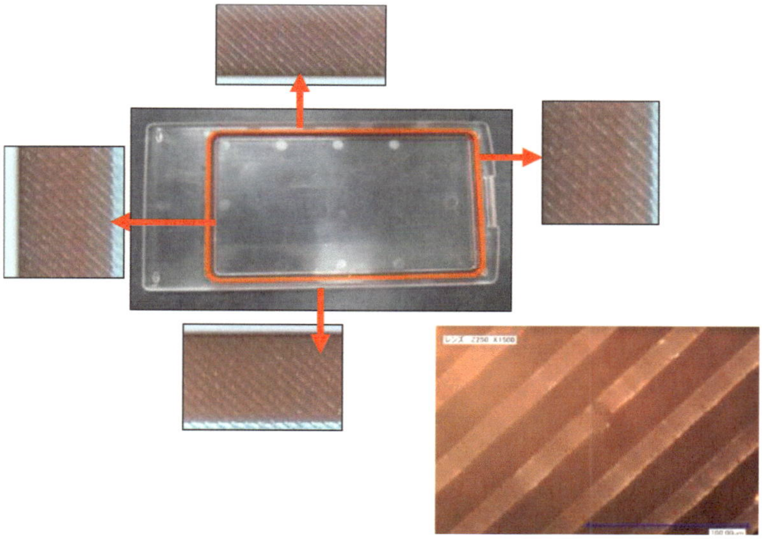

Figure 22.
A mobile phone PC-model joined with the silicone rubber through the microtexture with the width of 20 mm on the interface.

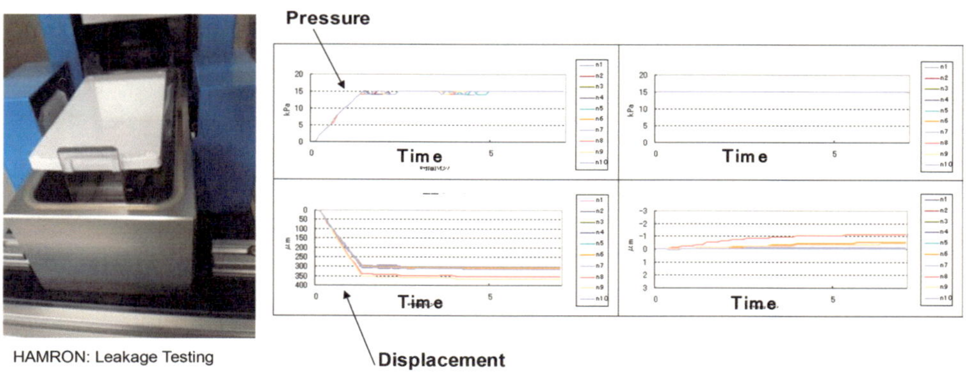

Figure 23.
Waterproof test to demonstrate the integrity of mobile phones under the pressure of 15 kPa.

in situ joining of silicone rubber ring onto the PC-cover case via the microbump textures on PC. **Figure 22** depicts the mobile phone model, fabricated in the above procedure. Each interface between the PC-cover case and silicone rubber has microbump textures. The Hamron leakage testing is employed to perform the waterproof test; e.g., this test aims at the quality check of significant deformation by small leaks under the applied pressure for 5 min. This model is dipped into a water pool, pressurized up to 15 kPa and held for 5 min. As shown in **Figure 23**, the PC-cover case deforms by pressuring it up to 15 kPa; no further deformation is detected during the holding duration. This demonstrates the perfect waterproof on the jointed interface with aid of microbump textures.

7. Surface property control by laser nano-/microtexturing

Superhydrophilicity and superhydrophobicity have grown up as a key surface engineering to keep clean and fresh surface of products and to control the liquid flow on the product surfaces. The oxide-glass lens as well as metallic-glass, optical elements are a typical targeting product to have their surface hydrophilic or superhydrophilic for liquid film formation, and to have it hydrophobic or superhydrophobic for well-defined water repellency [25]. The high energy surface had higher attractive capacity to other material atoms and molecules; those are adherent to each other to form a wet film on the surface. While, the low energy surface had lower attractive capacity to other material atoms and molecules; those are isolated from each other to form the droplets on the surface.

There are two modifications to control this surface state; e.g., the chemical and physical treatments. The chemical treatment is a general tool to modify the surface condition; e.g., fluorine-based coating increases the contact angle up to 130–150° in [26]. On the other hand, the idea of lotus effect has been discussed as a physical approach to form hydrophobic surface [27]. This lotus effect works in nature since the water droplets are supported by the air gap through the fine fibrous lotus leaf; this idea suggests that wettability might be widely controlled by the micro-/nanotexturing [28]. As has been reported in [29–31], the femtosecond laser micro-/nanotexturing methods have been developed to tune the surface wettability from superhydrophilic to superhydrophobic states. In particular, the micro-/submicro textures are formed on any materials by the laser-induced periodic surface structuring (LIPSS), where the incident and reflected lights have interaction with the scattered and diffracted lights at the vicinity of surface roughness [32]. Among several approaches to design this LIPSS, the authors proposed the micro-/submicrotexturing design by LIPSS with the use of fundamental wavelets and high-frequency ripples [33, 34]. Here, LIPSS is formed onto the AISI304 stainless steel substrates by using the femtosecond laser texturing. Both the superhydrophilic and superhydrophobic surfaces can be formed by the present laser nano-/microtexturing. The geometric effect of surface geometry on the superhydrophobicity is discussed to optimize the laser surface profile control.

7.1 LIPSS by femtosecond laser texturing

With reduction of the pulse duration, the optical interaction with irradiated materials localizes in the wavelength range. When irradiating the materials in the fundamental mode, this interaction field is limited within the submicrometer range. LIPSS is a typical local interaction, occurring at the site of material surface roughness in the order of micrometer. **Figure 24** depicts the LIPSS formed on the austenitic stainless steel type 304 by the present femtosecond laser texturing. Nanotexturing alignment angulates itself across the microtexture in **Figure 24a**

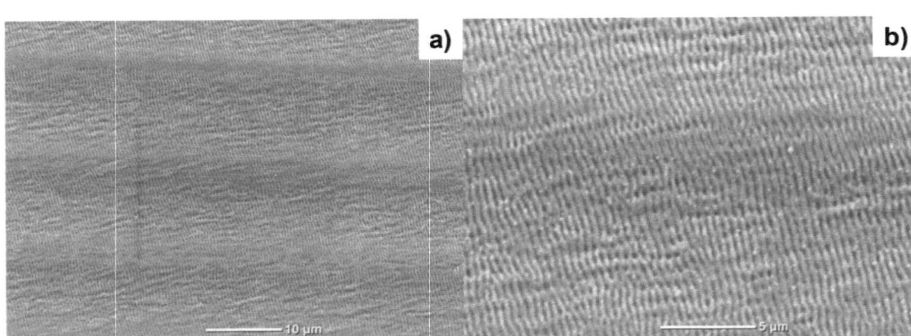

Figure 24.
LIPSS formed on AISI304 substrate surface by the present femtosecond laser texturing. (a) Microtextured angulation and nanotextures and (b) fine alignment of nanotextures.

since optical interaction is affected by the surface profile in micrometer range. As shown in **Figure 24b**, the spatial periodicity of these nanotextures is constant by 250 nm. This reveals that fine nanotextures with constant periodicity are formed on the metallic surface by the femtosecond laser treatment.

7.2 Superhydrophilic surface formation

After the classical theory on the surface wettability [35], the hydrophilic or the hydrophobic surfaces are modified to have superhydrophilic or superhydrophobic states, respectively. This is because the geometric item works to decrease the contact angle for hydrophilic surface or to increase it for hydrophobic one. **Figure 25** depicts the wettability of nanotextured AISI304 surface by the femtosecond laser surface modification. The measured contact angle reaches down to 8°; it is superhydrophilic. This reveals that the classical theory is true to describe the geometric nanotexture effect on the contact angle when the spatial periodicity of nanotextures works as a major geometric item in surface quality.

7.3 Superhydrophobic surface formation

In addition to the nanotexturing surface modification, the microtexturing angulation is taken into account as the geometric item. AISI304 stainless steel sheets with the size of 25 × 25 × 3 mm^3 are nano-/microtextured to investigate the change of surface wettability by this processing. **Figure 26** compares the droplets swelling on the specimen before and after this micro/submicrolaser texturing. The contact angle of pure water on the bare stainless steels is 70–75°, corresponding to the normal wettability of metals [36]. Through the present texturing, the contact angle

Figure 25.
Modification of hydrophilic surface to have superhydrophilic state by laser nanotexturing.

Figure 26.
Modification of wettability on the AISI304 substrate from the original hydrophilic state to the superhydrophobic one by laser nano-/microtexturing.

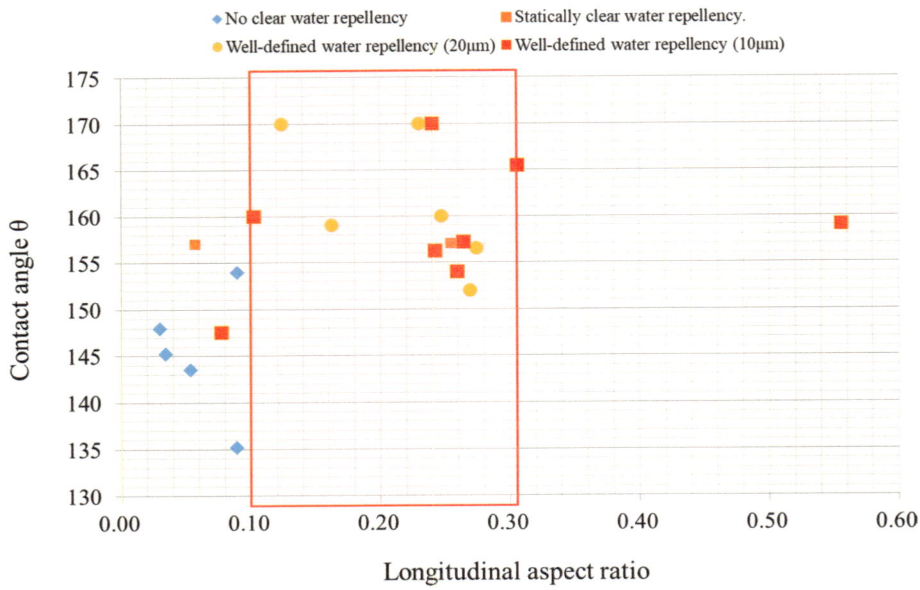

Figure 27.
Effect of the longitudinal aspect ratio on the measured contact angle among 30 nano-/microtextured AISI304 substrates.

increases up to 156°. This proves that nano-/microlaser texturing provides a tool to modify the wettability of stainless steel surfaces from hydrophobic to superhydrophobic state. This finding is completely against the classical theory; if more geometric items are put into laser texturing, the material surface quality can be widely controlled by geometric design.

7.4 Optimization of surface geometric configuration

There are two geometric items affecting on the surface property; the fractal dimension and the aspect ratio for nanotextures [37]. The former influences on the complexity of surface geometry; the latter, on the local angulation of geometry. Thirty AISI304 stainless steel sheets with the size of 10 × 10 × 0.1 t mm^3 are laser nano-/microtextured to investigate the effect of microtexture pitch and height on the measured wettability. **Figure 27** describes the relationship between the aspect ratio of nanotexture width to height on the measured contact angle. When this aspect ratio is less than 0.1 or more than 0.3, almost all measured contact angles are less than 155°; the micro-/submicrotextured AISI304 specimens are only hydrophobic. Higher contact angle up to 170° is attained when tuning this aspect ratio between 0.2 and 0.3; e. g., when using the microtextures with the width of 20 μm, their height might well be 2–6 μm. This implies that local angulation of surface geometry has significant influence on the controllability of hydrophobicity.

8. Friction control of tools by laser microtexturing

Under the strong demand for reduction of environmental burdens in manufacturing, every productive line must be energy-saving and highly material-efficient with less emission to environments [38]. In past, the huge amount of lubricating oils has been utilized to reduce the friction and wear not only in automobile industries but also in machining, metal forming, and so on [39]. In order to reduce this amount down to the minimum quantity, the contact surface of mechanical parts and tool surfaces are microtextured to reduce the friction coefficient and wear rate under minimum quantity lubrication (MQL) [40]. Microdimples on the working interfaces and surfaces play as a lubricating oil pocket to form a thin lubricating oil film on the interface between sliding parts and between work materials and tools [41]. The depth profile of each microdimple reflects on the local pressure distribution; this interfacial lubricating film works as a pressure boundary to support the sufficient film thickness to lubrication under MQL [42]. In addition, these microdimples work as a reservoir to store the wear debris of work materials and tool chips during the semidry machining and metal forming [43]. Here, the microdimples are formed by the picosecond laser texturing onto the dies and tools. The pin-on-ball method is employed to evaluate on the reduction of friction for the microdimpled die. The normal milling test is also utilized to describe the effect of microdimpled cutting tool on the reduction of tool wear.

8.1 Laser microtexturing of dimples

The picosecond laser microtexturing is employed to form the circular microdimples onto the AISI420 stainless steel dies, with the diameter of 50 and 100 μm and the depth of 10 μm in the regular lattice alignment with the pitch of 100 and 200 μm, respectively, for tribotesting. While, the isosceles triangular microdimples with the bottom edge of 155 μm, the height of 80 μm and the depth of 5 μm are machined onto the WC (Co) cutting tools in the zigzag alignment. **Figure 28** depicts these microdimpled specimens and tool together with the SEM-image and three-dimensional profile of microdimples.

8.2 Tribotesting

The pin-on-ball testing is employed to measure the time evolution of frictional force under the constant normal load. In this testing, the counter material ball is on contact with the die material under the applied normal weight as depicted in **Figure 29**. The frictional force is directly measured by load sensor attached to the arm. In the following tests, SUJ2 hard balls are utilized as a counter material. The friction coefficient is calculated by division of the measured friction force to the applied normal load. **Figure 30** depicts the transients of friction coefficient with increasing

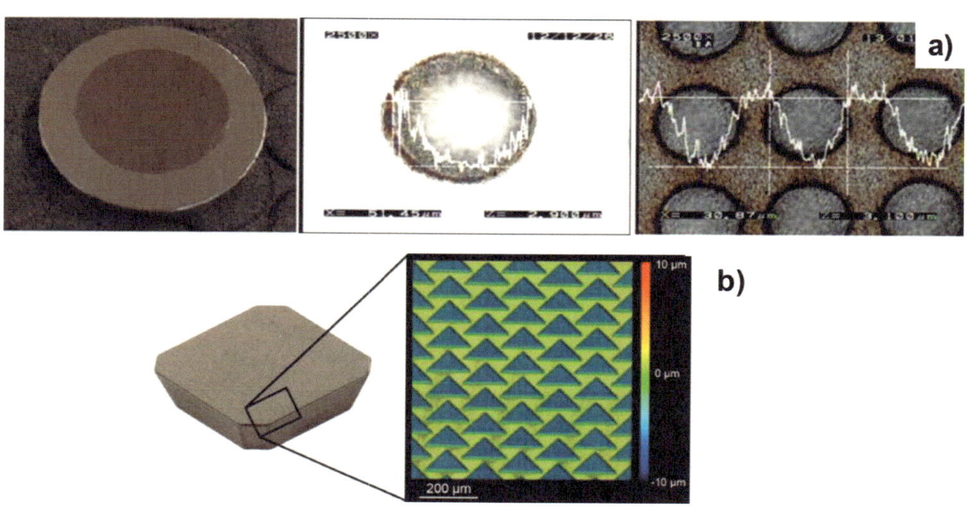

Figure 28.
Laser microdimple texturing. (a) Microdimpled stainless steel die and (b) microdimpled WC (Co) cutting tool.

the sliding distance for three die specimens; e.g., a bare AISI420 die without microdimples, and two microdimpled dies with the microdimple diameter (D) of 100 μm and its pitch (p) of 200 μm and with D = 50 μm and p = 100 μm, respectively. In case of bare die, the friction coefficient increases monotonically with sliding distance up to 0.15. When using the microdimpled die with D = 50 μm and p = 100 μm, lower friction coefficient than 0.07 is preserved during this tribotesting.

8.3 Machining test

When milling the aluminum alloys by WC (Co) tools, the tool face is inevitably subjected to adhesion of work material. Microtexturing into the tool face enables to reduce this adhesion by storing the wear debris and cutting chips into these pockets on it. In this experiment, AA5052 aluminum alloy is employed as a work material for normal milling with use of the bare WC (Co) and microtextured one as shown in **Figure 28b**. **Figure 31** compares the adhesion process of work material onto the tool face at the milling distance (L) of 900 and 1800 m, respectively, between the bare and microdimpled tools. Without microdimples, the adhesive area and thickness of work materials onto the tool face enlarges with increasing L; e.g., when

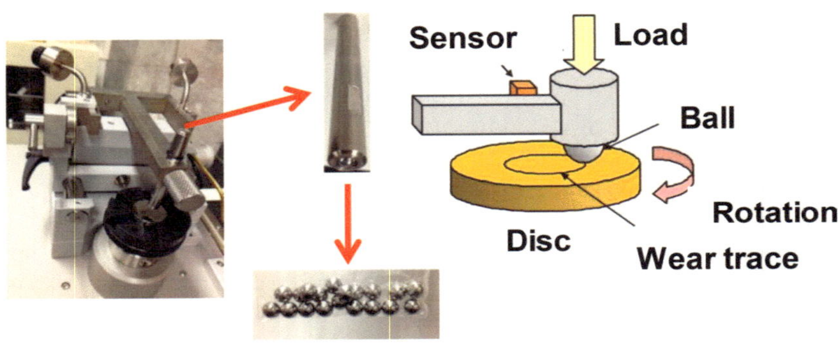

Figure 29.
The ball-on-disc method for measurement of friction coefficient during the sliding conditions.

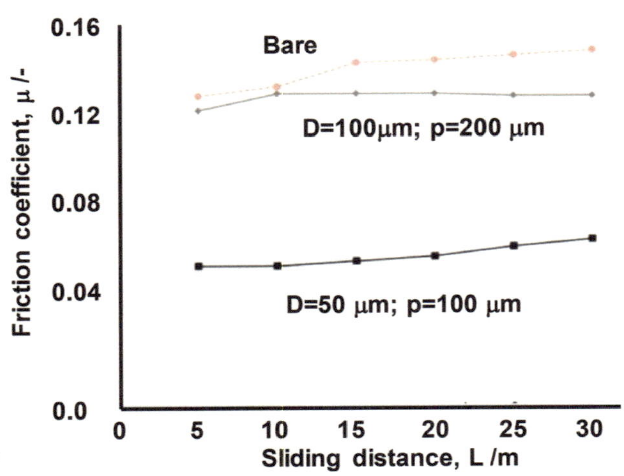

Figure 30.
Variation of the friction coefficient with increasing the sliding distance for three cases; the bare die without microdimples, and the microdimpled dies with D = 100 μm and p = 200 μm and D = 50 μm and p = 100 μm, respectively.

L = 1800 m, nearly the whole face is covered by these work adhesives with their film thickness of 10 μm around the tool edge. On the other hand, little adhesion to microdimpled face is noted even after milling up to 1800 m. This significant reduction of adhesion by microtexturing comes from the storing mechanism where the wear debris and cutting chips are reserved into each microdimple. This reduction of adhesion influences on the cutting force; e.g., the cutting force becomes relatively insensitive to cutting distance when using these microtextured tools.

8.4 Applications

Low friction and wear is indispensable for most of automotive parts and manufacturing tools. They have curved surfaces, the friction coefficient of which must be reduced to save the energy waste and to improve the fuel efficiency. In particular,

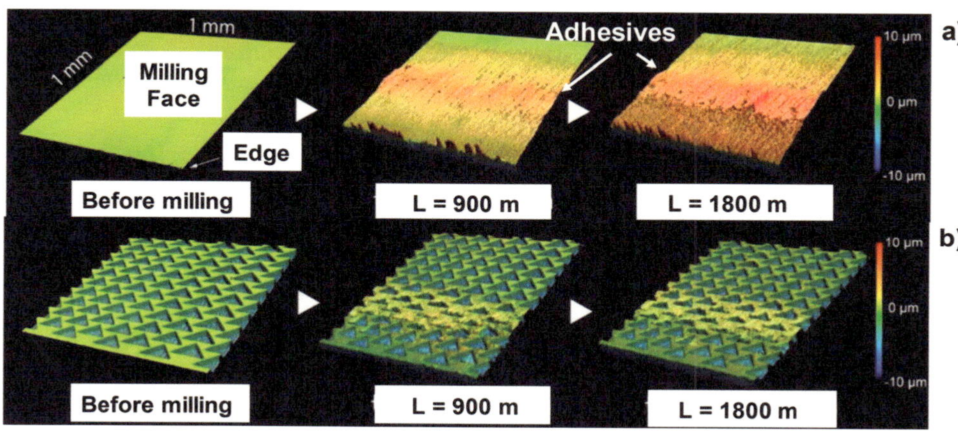

Figure 31.
Comparison of work material adhesion to tool face with increasing the milling distance between the bare and microdimpled WC (Co) tools.

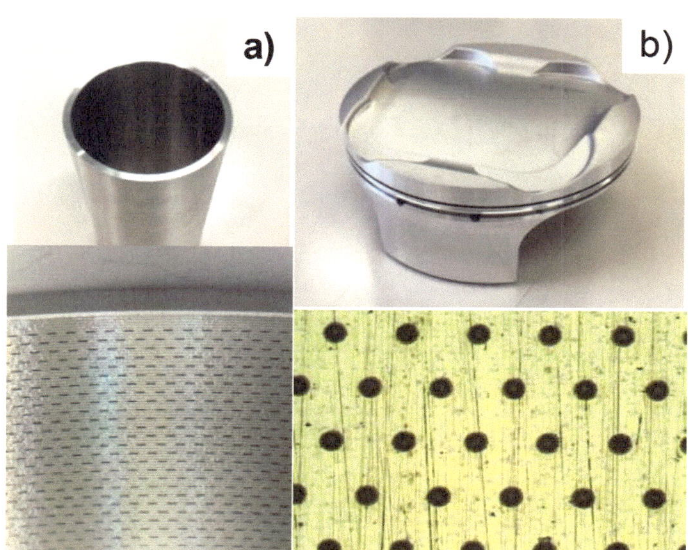

Figure 32.
Microtexturing into the inner surfaces of automotive parts. (a) Microtextured piston cylinder and (b) microtextured piston skirt.

the piston cylinder as well as piston skirt are important sliding-part. **Figure 32a** shows the microdimpled AISI316L inner surface of cylinder with the size of 30 × 500 μm^2 and the depth of 5 μm in the pitch of 1 mm in the circumferential direction and 0.5 mm along the length. This wedge-shaped microdimples improve the fuel efficiency significantly. The AA7075 piston skirt is also microdimpled to have circular dimples with the diameter of 30 μm, the depth of 3 μm and the pitch of 120 μm, respectively, as shown in **Figure 32b**.

9. Discussion

The spatial resolution in this laser machining is first discussed to find out the way to improve its dimensional accuracy. Through the practical survey on the micromachining and texturing into curved surfaces, the feasible applications are understood to search for bio-medical laser processing. In particular, future trend of fast-rate laser technologies is discussed for further improvement of micromachining.

9.1 Spatial resolution in laser micromachining

Laser drilling of circular holes is employed as a benchmark test to discuss the dimensional accuracy of 25 × 25 holes in square structure with the diameter of 30 mm and the pitch of 50 mm, as depicted in **Figure 33a**. Silicon nitride plate with the thickness of 125 mm is used as a work material. **Figure 33b** and **c** shows the

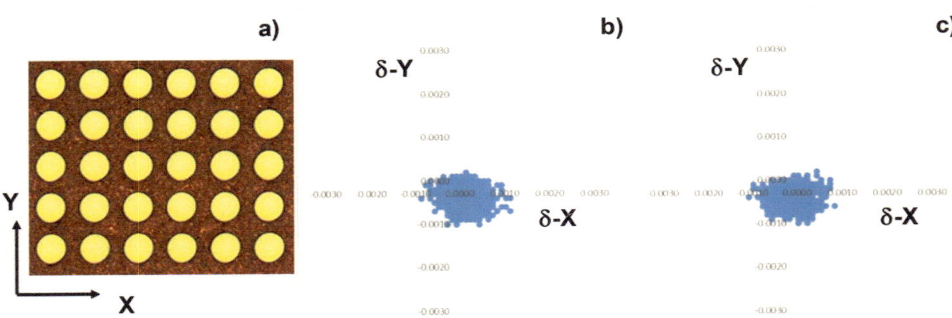

Figure 33.
Benchmark test to investigate the dimensional accuracy in the laser microdrilling. (a) Test-drilling, (b) deviation map, measured at the inlet, and (c) deviation map, measured at the outlet.

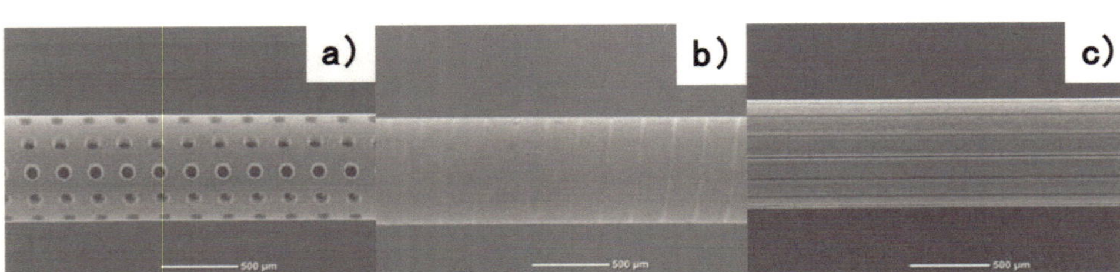

Figure 34.
Microtexturing into the metallic tube. (a) Microdrilling of holes with the diameter of 100 μm into a AISI304 stainless steel pipe with D_{out} = 0.7 mm and D_{IN} = 0.58 mm, (b) microgrooving of shallow grooves with the width of 25 μm and the depth of 3 μm into the same pipe as (a), and (c) microgrooving of lateral grooves with the depth of 30 μm into the same pipe as (a).

X- and Y-deviation maps at the inlet diameter and outlet diameter for 625 holes. Since both maps are nearly coincident to each other, the straightness and circularity are preserved to be within 1/125 μm ~ 0.7°, and within 2 μm, respectively. The spatial resolution of hole diameter is within 2.5 μm in the 2σ-reliability.

9.2 Laser micromachining into curved surfaces

Without specially designed jigs and fixtures, both the micromilling and micro-EDM are difficult or nearly impossible to microdrill the holes and grooves. Laser microdrilling has little constraint in the manufacturing setup; it is readily applied to make direct drilling. AISI304 stainless steel pipe with its outer diameter (D_{out}) of 0.7 mm and its inner diameter (D_{in}) of 0.58 mm is employed as a work to make microdrilling the holes and grooves. **Figure 34** shows three microtexturing cases; e.g., a microdrilled pipe, a spiral-grooved pipe, and a laterally grooved pipe. The designed textures can be accommodated to the curved surfaces by this laser microdrilling.

Another feature of laser microtexturing is developed by changing the beam control. A thin spring is structured from a pipe in **Figure 35a**. A wide slit is structured into a pipe as depicted in **Figure 35b**. Any shaped short-cuts are equipped into a pipe as shown in **Figure 35c**. This suggests that complex microstructure can be built in the micromachine and micromember.

9.3 Laser micropart formation from pipes

Let us discuss how to make laser-structuring a micropart from commercial components. A polylactic acid (PLA) pipe is employed as a starting component to fabricate the PLA-stents for medical usage. **Figure 35** depicts three PLA-stents fabricated from the same PLA-pipe by the laser microtexturing. These three can be selectively made from PLA-pipe only by varying the slit width (Ws) to be 154,

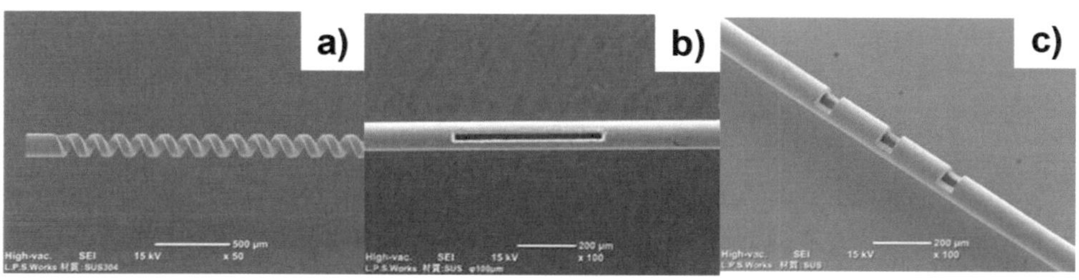

Figure 35.
Fabrication of the geometrically functionalized parts by laser microtexturing. (a) Structuring a spring with the pitch of 150 μm from thin brass pipe with D_{out} = 160 μm and D_{in} = 80 μm, (b) structuring a slit with the size of 30 × 500 μm^2 into AISI304 stainless steel pipe with D_{out} = 100 μm, and (c) fabrication of short-cuts with the size of 65 × 50 μm^2 from the same pipe as (b).

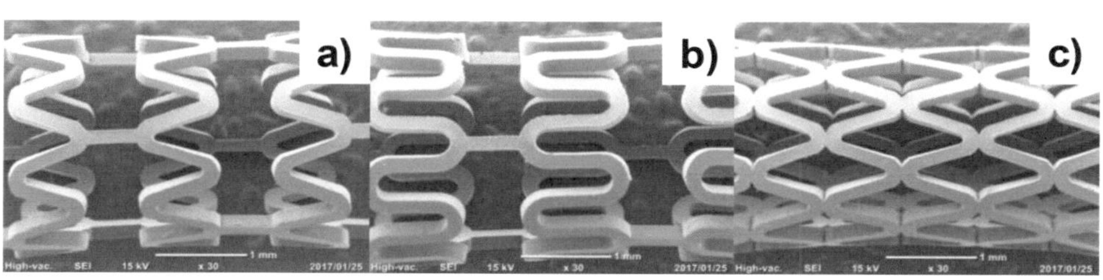

Figure 36.
Laser micropart formation of stents from PLA pipe with D_{out} = 2.55 mm, D_{in} = 2.20 mm, and the length of 25 m. (a) Ws = 154 μm, (b) Ws = 156 μm, and (c) Ws = 160 μm.

156 and 160 micro-meter, respectively. The topological geometry of stents can be designed and fabricated for each medical treatment by tuning the laser microtexturing parameters as shown in **Figure 36**.

9.4 Future trends in laser micromanufacturing

Various geometric transformations can be realized only by the laser processing such as the micromachining, microtexturing, and microstructuring in the above. Through the fusion of other manufacturing treatments with the laser processing, further advancement is expected to propose the innovative procedures. With combination of laser nano-/microtexturing with laser polishing, the surface property is selectively controlled to be superhydrophilic or superhydrophobic by tuning the LIPSS-conditions. With combination of laser microtexturing with the mechanical milling, a multimaterial part as well as a structural member with large area can be functionalized as a complex-shaped part or as a functionalized component.

Among the engineering issues related to ultrashort pulse laser processing, how to put the fast-rate microtexturing into practice is one of the important targets. In addition to increase of repetition frequency in laser oscillation, new optical control must be developed to transform the spatial geometry and topology in shape into time sequence of scanning in beam technology.

10. Conclusion

The picosecond and femtosecond laser processing is designed to be tools for advanced manufacturing; laser microdrilling, laser microtexturing, laser nano-/microtexturing, laser microstructuring, and so on. The dimensional accuracy, the spatial resolution as well as the circularity approaches to 1 μm or less than; every micropart, every microstructure, and every microtexture is fabricated in the product size of 10 to 100 μm range. Most of engineering issues related to surface and interface are well defined in this laser processing to find an optimum solution to each problem. Reduction of friction and wear in tools and works is attained by microtexturing onto the tool and part surfaces. Reliable joining between dissimilar materials and parts is put into practice by chemical adhesion with aid of microtextures on their interface. Surface and interface properties are also controllable by optimization of nano-/microtextures.

Sustainable manufacturing requires for the well-designed processing to support the efficient circulation of products, parts, and materials in addition to recycle and reuse of second hands. Laser micromachining is useful to prolong the tool life, to revise the product surfaces and interfaces for multiple use and to assist the multi-materialization for second-hand products and parts.

Further research and development on the unknown features of laser processing is necessary to advance new steps in innovative technology and medical engineering to further improve the sustainability in future society.

Acknowledgements

The authors would like to express their gratitude to Mr. T. Hasegawa, T. Miyagawa (SIT), Dr. K. Wasa (TecDia, Co. Ltd.), Mr. T. Omata, and Mr. K. Sanbongi (LPS-works, Co. Ltd.) for their help in experiments. The present study was financially support in part by the METI with Supporting-Industry Projects in Japanese Government from 2010 to 2017.

DOI: http://dx.doi.org/10.5772/intechopen.83741

Conflict of interest

The authors declared no conflict of interest.

Author details

Tatsuhiko Aizawa[1]* and Tadahiko Inohara[2]

1 Surface Engineering Design Laboratory, Shibaura Institute of Technology, Tokyo, Japan

2 LPS Works, Co. Ltd., Tokyo, Japan

*Address all correspondence to: taizawa@sic.shibaura-it.ac.jp

References

[1] Avanish D. Laser beam machining—A review. International Journal of Machine Tools and Manufacture. 2008;**48**:609-628

[2] Davim P. Nontraditional Machining Processes: Research Advances. New York: Springer; 2013

[3] Schmid K. Manufacturing Processes for Engineering Materials. 5th ed. New Jersy: Prentice Hall; 2008

[4] Aizawa T, Iohara T. Japan Patent. 2011-212046; 2011

[5] Gaertner E, Polise V, Tagliaferri F, Palumno B. Laser micro machining of alumina by a picosecond laser. Journal of Laser Micro/Nanoengineering. 2018;**13**:76-84

[6] Amer MS, Ei-Ashry MA, Dosser LR, Hix KE, Maguire JF, Irwin B. Femtosecond bversus nanosecond laser machining: Comparison of induced stresses and structural changes in silicon wafers. Applied Surface Science. 2005;**242**:162-167

[7] Nasrollahi V, Penchev P, Dimov SS. A new laser drilling method for producing high aspect ratio micro holes. In: Proc. 11th 4M Conf. 2016. pp. 15-18

[8] Aizawa T, Inohara T. Micro-texturing onto glassy carbon substrates by multi-axially controlled pico-second laser machining. In: Proc. 7th ICOMM. 2012. pp. 66-73

[9] Aizawa T, Inohara T. Multi-dimensional micro-patterning onto ceramics by pico-second laser machining. Research Reports of Shibaura Institute of Technology. Natural Sciences and Engineering. 2012;**56**(1):17-26

[10] Tokura K. Machining process. Japanese textbook for mechanical engineering. The Japan Society of Mechanical Engineers. 2006;**5**:111

[11] Doering S, Richter S, Molte S, Tuennermann A. In-situ observation of the hole formation during deep drilling with ultra-short pulses. Proceedings of SPIE. 2011;**7925**:792517:1-792517:8

[12] Doering S, Richter S, Nolte S, Tuennermann A. In-situ imaging of hole shape evolution in ultra-short pulse drilling. Optics Express. 2010;**18**:20395-20400

[13] Ashkenasi D, Mueller N, Kaszemeikat T, Illing G. Advanced laser micro machining using a novel trepanning system. Journal of Laser Micro/Nanoengineering. 2011;**6**(1):1-5

[14] Aizawa T, Inohara T. Pico-second laser drilling of high-aspect ratio through-holes with and without tapering. In: Proc. 1st WCMNM. Vol. 103. 2017. pp. 1-2

[15] Aizawa T, Wasa K, Tamagaki H. A DLC-punch array to fabricate the micro-textured aluminum sheet for boiling heat transfer control. Micromachines. 2018;**9**:147-1-147-10

[16] Denkena B, Koehler J, Kaestner J. Efficient machining of micro-dimples for friction reduction. In: Proc. 7th ICOMM. 2012. pp. 85-89

[17] Jiang Y, Zhao WS, Kang XM, Gu L. Adaptive control for micro-hole EDM process with wavelet transform detecting method. In: Proc. 6th ICOMM. 2011. pp. 207-211

[18] Aizawa T, Fukuda T. Oxygen plasma etching of diamond-like carbon coated mold-die for micro-texturing. Surface and Coating Technology. 2013;**215**:364-368

[19] Aizawa T, Itoh K, Inohara T. Imprinting of patterns onto polymers and oxide-glasses via fine micro-stamping. In: Proc. 6th ICOMM. 2010. pp. 77-82

[20] Aizawa T. Micro-texturing onto amorphous carbon materials as a mold-die for micro-forming. Applied Mechanics and Materials. 2013;**289**:23-37

[21] Balden A. Adhesively bonded joints and repairs in metallic alloy, polymers and composite materials: Adhesives, adhesion theories and surface pretreatment. Journal of Materials Science. 2004;**39**:1-49

[22] Available from: http://www.shinetsu.co.jp/ [Accessed: Nov 30, 2018]

[23] Aizawa T, Satoh S, Yamaguchi T. Micro-texturing design for joining between polymer components. In: Proc. 9th ICOMM. Vol. 15. 2014. pp. 1-8

[24] Aizawa T, Satoh S, Yamaguchi T. Joining of polymer systems by micro-texturing toward perfect water-proof. Research Report, SIT. 2014;**58**:1-10

[25] Packham DE. Surface energy, surface topography and adhesion. International Journal of Adhesion and Adhesives. 2003;**23**:437-448

[26] Nakamae K. How to form the hydrophobic and super-hydrophobic surfaces in industries. Industrial Materials. 1996;**44**:26-30

[27] Cheng YT, Rodak DE. Is the lotus leaf superhydrophobic? Applied Physics Letters. 2005;**86**:144101

[28] Aizawa T. Micro-texturing for tribology and surface engineering in manufacturing processes. In: Proceedings ISAST IV. 2016. pp. 1-16

[29] Hoehm S, Rosenfeld A, Krueger J, Bonse J. Femtosecond laser-induced periodic surface structures on silica. Journal of Applied Physics. 2012;**1121**:0149010-0149019

[30] Orazi L, Gnilitskyi I, Serro AP. Laser nanopatterning for wettability applications. Journal of Micro and Nano-Manufacturing. 2017;**5**:0210081-0210088

[31] Kietziga AM, Mirvakilia MN, Kamalb S, Englezosa P, Hatzikiriakos SG. Laser-patterned super-hydrophobic pure metallic substrates: Cassie to wenzel wetting transitions. Journal of Adhesion Science and Technology. 2011;**25**:2789-2809

[32] Gečys P, Vinčiūnas A, Gedvilas M, Kasparaitis A, Lazdinas R, Račiukaitis G. Ripple formation by femtosecond laser pulses for enhanced absorptance of stainless steel. Journal of Laser Micro/Nanoengineering. 2015;**10**:129-133

[33] Hasegawa T, Aizawa T, Inohara T, Wasa K. Fabrication and control of super-hydrophobic surfaces by the femto-second laser machining. In: Proc. 1st World Congress on Micro and Nano Manufacturing. 2017. pp. 381-382

[34] Hasegawa T, Aizawa T, Inohara T, Wasa K. Engineering design for super-hydrophobic surfaces via the femto-second laser machining. In: Proc. 10th Asian Workshop on Micro/Nano Forming Technology. 2017. pp. 33-34

[35] Kawase T. Super-hydrophobic surface. In: SEM'I GAKKAI. Vol. 65. 2009. pp. 200-207

[36] Kam DH, Bhattacharya S, Mazumder J. Control of the wetting properties of an AISI 316L stainless steel surface by femtosecond laser induced surface modification. Journal of Micromechanics and Microengineering. 2012;**22**:1-6

[37] Hasegawa T, Aizawa T, Inohara T, Wasa K. Hot mold stamping of optical

plastics and glasses with transcription of super-hydrophobic surfaces. Procedia Manufacturing. 2018;**13**:1437-1444

[38] Allwood JM, Cullen JM. Sustainable Materials. Cambridge: UIT; 2012

[39] Kataoka S. Influence of lubricants on global environment. Journal of the Japan Society for Technology of Plasticity. 2005;**46**:4-10

[40] Czichos H, Habig K-H. Tribologie-Handbuch (Tribology handbook). 2nd ed. Wiesbaden: Vieweg Verlag; 2003

[41] Aizawa T, Morita H. Dry progressive stamping of copper-alloy snaps by the plasma nitrided punches. Materials Science Forum. 2018;**920**:28-33

[42] Aizawa T, Morita H. Tribological design and engineering in surface treatment for semi-dry and dry stamping. In: Proc. ICTMP. 2016. pp. 14-28

[43] Sugihara T, Enomoto T. Performance of cutting tools with dimple textured surfaces: Comparative study of different texture patterns. Journal of the International Societies for Precision Engineering and Nanotechnology. 2017;**49**:52-60

Chapter 3

Silicon-Based Micromachining Process for Flexible Electronics

Jiye Yang and Tao Wu

Abstract

In this chapter, we introduce silicon-based micromachining process and devices for flexible electronics application. Silicon-based flexible electronics have the unique advantage over other polymer-based process that leverage the traditional standard CMOS process and can be integrated with scalable IC technology. While integrating with CMOS process, special considerations must be taken into account, such as release process, transfer process, and process integration, in order to produce silicon-based flexible electronics. Several efforts and process developments will be illustrated in this chapter with the highlights of imager and wearable electronics application.

Keywords: silicon-based flexible electronics, silicon-on-insulator (SOI), deep reactive-ion etching (DRIE), release process, chemical mechanical polishing (CMP), atomic layer deposition (ALD), complementary metal-oxide semiconductor (CMOS), through-silicon via (TSV), frontside-release process, backside-release process

1. Introduction

Over the past decade, an enthusiastic pursuit for flexible electronics, employing both organic and inorganic semiconductor materials, with continuously improved performance has been observed [1]. The material like polymer, carbon nanotubes (CNTs), and silicon (Si) membrane are popular candidates for flexible electronics. Compared with other materials, monocrystalline Si nanomembrane released from silicon-on-insulator (SOI) emerges as one of the best choices due to its high carrier mobility, commercial availability at relatively lower cost, and mature fabrication techniques [2]. Recently, nanostructured silicon has been widely used to produce flexible electronic devices like flexible solar cells, thermal electricity, and piezoelectric generators.

Functional part of flexible electronics based on silicon can be fabricated in a standard complementary metal-oxide semiconductor (CMOS) technology. A standard CMOS technology includes photolithography, etch, deposition, and doping. Moreover, frontside- and backside-release process, transfer process, and bonding process for flexible substrate are developed to generate flexible silicon membrane with functional part. For frontside-release process, deep reactive-ion etching (DRIE), buffered oxide etcher (BOE), or xenon difluoride (XeF_2) etching are used to release membrane structures, while for backside-release process, lapping, chemical mechanical polishing (CMP), or XeF_2 etching are employed to thin the Si substrate. After fabricating thin Si membranes with functional devices, special transfer process is up required to stick released devices on a flexible substrate like PDMS or Kapton® tape. After the release process, the released Si membrane is transferred and bonded to a flexible substrate (**Figure 1**).

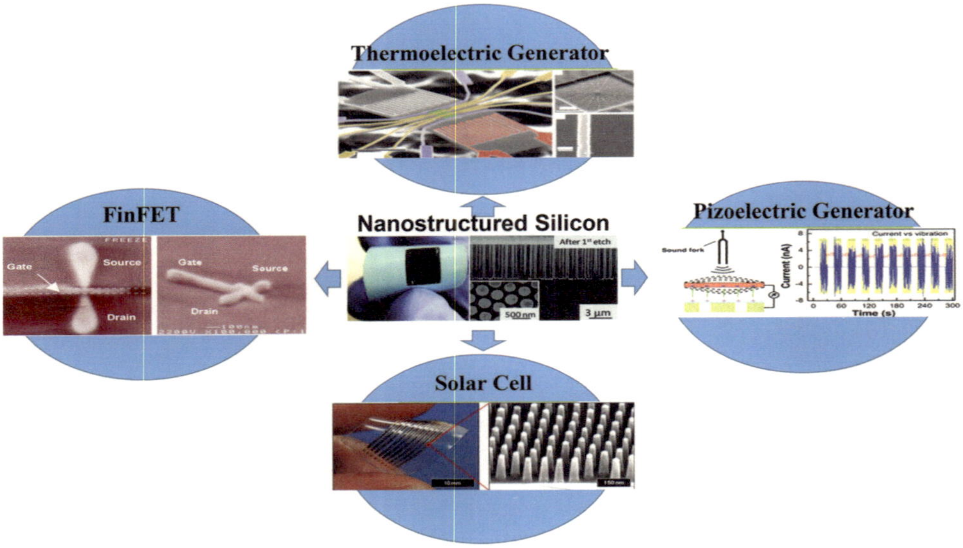

Figure 1.
Schematic diagram of Si-based nanostructures that are served as flexible thermoelectric generator, solar cells, ICs, and piezoelectric generators. As shown in the middle, Si nanowire array on plastic substrate and its cross section under scanning electron microscopy (SEM) are flexible. Images at the bottom: Reproduced with permission [3]. Copyright 2012, American Chemical Society. Images in the middle: Reproduced with permission [4]. Copyright 2011, American Chemical Society. Images in the middle right: Reproduced with permission [5]. Copyright 2011, American Chemical Society. Images in the middle left: Reproduced with permission [6]. Copyright 2011, American Chemical Society. Images at the top: Reproduced with permission [7]. Copyright 2008, nature publishing group [8].

2. Silicon-based micromachining process

Many researchers have demonstrated flexible electronics based on polymers [9, 10] and CNT [11, 12]; however, the micromachining process with those materials is largely limited by the process temperature and compatible chemicals. Moreover, the devices are typically not scalable or almost impossible to integrate with current advanced IC technology. Compared to polymers and CNT, Si-based flexible electronics can employ the matured CMOS fabrication techniques such as photolithography, atomic layer deposition (ALD), physical vapor deposition (PVD), chemical vapor deposition (CVD), Hydrofluoric Acid (HF) etching, reactive-ion etching (RIE), etc. However, in order to produce flexible electronics using traditional Si-based COMS process, development of release process, transfer process, and bonding process are essential for the production of flexible thin silicon membrane. To realize flexibility the oft used processes are DRIE, XeF_2 dry etching, transfer through polymer stamp, process for bonding to flexible substrate, etc. On the whole, the fabrication of silicon-based flexible electronics consists of two major steps: the fabrication of functional part like photodiode, metal-oxide semiconductor field-effect transistor (MOSFET), fin field-effect transistor (FinFET), ferroelectric RAM (FeRAM), etc. and the thinning of device to realize the flexibility. In this chapter, micromachining processes are introduced and described.

2.1 Traditional silicon CMOS process

CMOS process is a standard process used to produce integrated circuits (ICs) and form electronic circuits and system in large scale. CMOS process involves various basic fabrication processes such as wafer manufacturing, oxidation, photolithography, doping, deposition, etching, and CMP.

2.1.1 Wafer manufacturing

Silicon wafers are produced from raw material sand by purifying and crystallizing. The purified silicon is held in molten state at about 1500°C, and after dipping a seed crystal into the melt, the silicon ingot can be produced by gradually extracting the rod. In addition, the silicon can be lightly doped by inserting doping material into the crucible. The fabricated silicon material is used to produce the CMOS device such as MOSFET and FinFET. **Figure 2a** shows the fabricated traditional bulk Si wafer. Nowadays the most advanced transistors are FinFET or fully depleted silicon-on-insulator (FD-SOI) planar transistor technology that is developed at the scale smaller than 25 nm. In the fabrication of FD-SOI transistor, instead of the traditional bulk Silicon wafer, the new more expensive material called SOI wafer is employed. The SOI wafer is fabricated by either separation by implantation of oxygen (SIMOX) process or Smart-Cut process [13]. **Figure 2b** shows SOI wafer for FD-SOI transistor. The thickness of the silicon film is in the ranges from 10 nm to 30 nm; while the standard thickness of BOX is approximately 145 nm and the thickness of ultra thin BOX ranges from 10 nm to 30 nm [14].

2.1.2 Oxidation

In the CMOS fabrication, silicon dioxide layer is used as an insulating material between different conducting layers or acts as a mask or protective layer against diffusion and high-energy ion implantation. The oxidation is performed by a chemical reaction between oxygen (dry oxidation) or water vapor (wet oxidation), and the silicon slice surface is heated in a high-temperature furnace at about 1000°C [15]. Dry oxidation is often used to produce thin and robust oxide layers, while wet oxidation is used to produce thicker and slightly porous layers.

2.1.3 Photolithography

With the help of mask, the photolithography is employed to create patterned layers of different materials on the silicon wafer. Photolithography involves several steps. At first, a photosensitive emulsion (photoresist) film is coated on wafer surface using spin coat. Following that, the wafer is exposed to a pattern of intense light with the help of mask. For positive photoresist (PR), the exposed regions are soluble in the developer, while for negative photoresist, the unexposed regions are soluble in the developer. Tetramethylammonium hydroxide (TMAH) is a widely used developer to remove unwanted photoresist regions. After development, the etching is performed to remove the unwanted regions that are not protected by photoresist. In projection systems, the resolution is limited by the wavelength of the light and the ability of the reduction lens system to capture enough diffraction orders from the illuminated

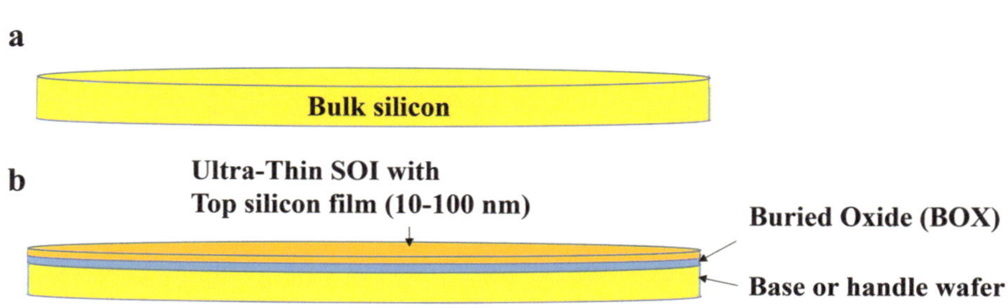

Figure 2.
(a) Traditional bulk silicon wafer. (b) FD-SOI starting wafer.

mask. Nowadays, the most advanced CMOS photolithography is at the scale of about 7 nm, but for flexible electronics photolithography at the scale of about 1 μm is enough for most application such as imager, temperature sensor, and humidity sensor.

2.1.4 Doping

Doping is used to produce electronic components such as diode and various transistors. After masking some area of the silicon surface, doping can be done in exposed regions. Doping can be performed by either diffusion method or ion implantation. There are two basic steps for diffusion method: predeposition and drive-in. In the predeposition step, the wafer is heated in a furnace to a certain temperature (about 1000°C), and carrier gas such as nitrogen and argon with the desired dopant such as phosphine PH_3 or diborane B_2H_6 flow to the silicon wafer. The diffusion of dopant atoms takes place onto the surface of the silicon, and in this step we can control the dose of dopant atoms. In the drive-in step, the wafer is heated in an inert atmosphere for few hours to distribute the atoms more uniformly and to a higher depth [15]. For ion implantation method, charged dopants (ions) are accelerated in an electric field and penetrated into the wafer. The penetration depth can be precisely controlled by reducing or increasing the voltage that needed to accelerate the ions. Following ion implantation, a drive-in step is also performed to achieve uniform distribution of the ions and increase the depth of penetration. **Figure 3** shows two phosphorous doping processes for SOI wafer, in which secondary ion mass spectrometry (SIMS) was used to analyze the doping profiles under two implantation conditions: one has energy/dose of 12 keV/1 × 10^{16} cm^{-3}, and the other has 150 keV/4 × 10^{15} cm^{-3} [1].

2.1.5 Deposition

For MOS Fabrication, various deposition methods are used to form conducting insulating and passivation layers with a variety of materials. There are three

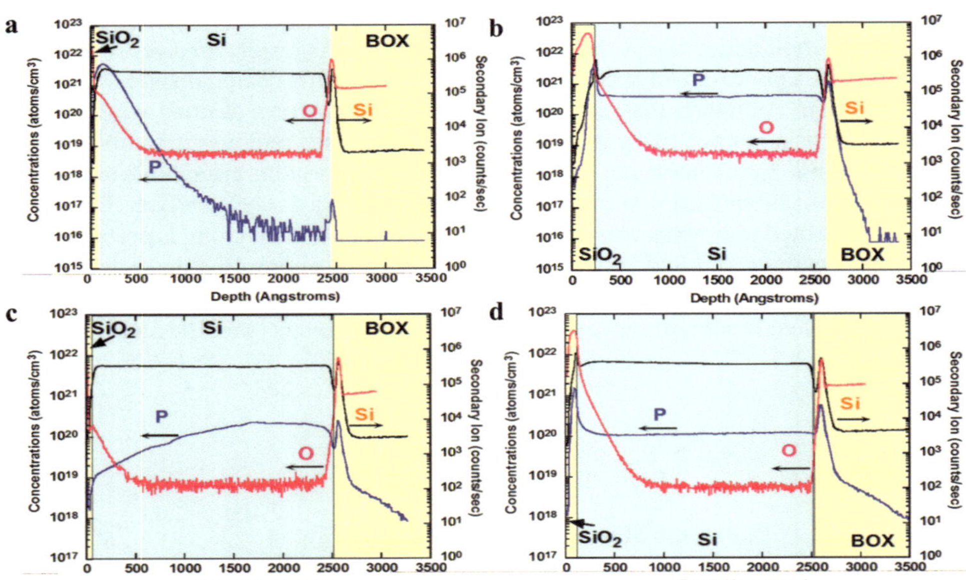

Figure 3.
SIMS results of phosphorous doping profiles of two implantation conditions: 12 keV/1 × 10^{16} cm^{-3} (a) before and (b) after annealing; 150 keV/4 × 10^{15} cm^{-3} (c) before and (d) after annealing. Reproduced with permission [1]. Copyright 2012, IOP.

main deposition processes: PVD, CVD, and ALD. CVD is widely used to deposit conducting layers such as polysilicon and insulating layers such as SiO_2 and Si_3N_4. ALD is used to deposit gate dielectrics with high-k material such as hafnium dioxide HfO_2 and tantalum pentoxide Ta_2O_5 that are necessary for FET at scale smaller than 25 nm. PVD is an established method of depositing metal contacts, barriers, and interconnects used in ICs [16]. In the advanced CMOS, a 3D stack chip structure is used to further improve the integration by using solder flip chip and through-silicon vias (TSVs). For the fabrication of TSVs, the depositions of passivation layer such as silicon nitride (SiN) and metal layer such as copper (Cu) are necessary. But some deposition processes are not more suitable for TSVs with aspect ratios more than 10:1; the capability of several deposition processes to coat the sidewalls of TSVs is limited as shown in **Figure 4**. Compared with ALD, the deposition coverage of CVD decreases below 20% for aspect ratios exceeding 10:1, and for aspect ratios larger than 2.5:1 the deposition coverage of PVD is already less than 20%. Moreover, molecular vapor deposition (MVD) is an alternative deposition process that is suitable for TSVs with aspect ratios larger than 10:1 [17].

2.1.6 Etching

Etching process is used to remove unwanted material and to create desired pattern. There are two types of etching methods: wet etching and dry etching. For wet etching, the wafer is immersed in a suitable etching solution, which can remove the exposed material leaving the material beneath the protective layer intact. For example, potassium hydroxide (KOH) is used to etch silicon, while hydrofluoric acid (HF) is used to etch SiO_2. In addition, the etching mask should not dissolve or at least be etched much slower in the etchant. For example, SiO_2 and Si_3N_4 can serve as mask for Si etching in KOH, while Si_3N_4 and metal are usually used as SiO_2 wet etching mask. Dry etching, usually called plasma etching or reactive-ion etching (RIE), is used to remove the materials by chemical reactions (using chemical reactive gases or plasma) and by purely physical methods (e.g., sputtering and ion beam-induced etching) or with a combination of both chemical reaction and physical bombardment (e.g., RIE). For instance, SF_6 and CF_4 can be utilized to etch silicon anisotropically, while XeF_2 etches silicon isotropically with pure chemical reaction. Depending on the selectivity and how much materials need to be etched, PR, SiO_2, or metal can be used as the mask for silicon etching [19, 20].

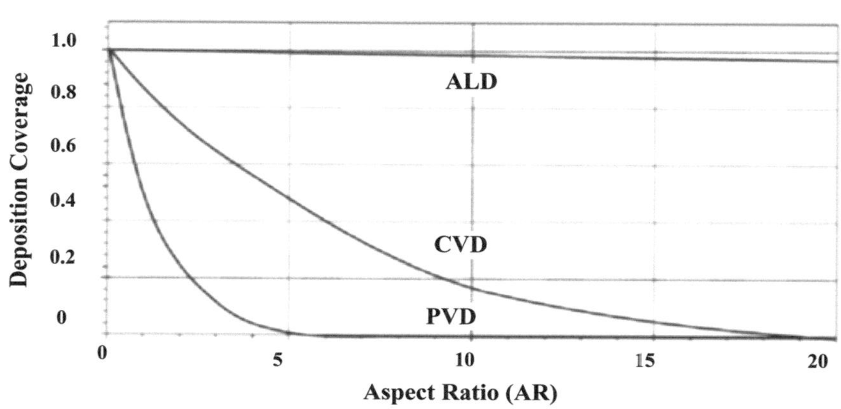

Figure 4.
Schematic graph of deposition coverage in comparison of PVD, CVD, and ALD deposition processes. Reproduced with permission [18]. Copyright 2016, springer international publishing Switzerland.

Etching can be isotropic or anisotropic and therefore can form different etching profiles. Isotropic etching has the same etch rate in all directions and, anisotropic etching has different etch rates in the lateral and vertical directions. For example, silicon can be etched anisotropically by using CF_4 or SF_6 and can be etched isotropically using XeF_2 or $HF:HNO_3:H_2O$.

2.1.7 Chemical mechanical polishing (CMP)

CMP process is a combination of mechanical and chemical actions, and it has been widely used to polish and thin silicon substrate. A CMP process could be significantly influenced by many factors such as abrasives, pH, and polishing temperature [21]. The schema of CMP tool and process to polish wafer are shown in **Figure 5** [22]. A wafer is firstly held by the polishing head using a vacuum and then the polishing head starts to rotate, resulting in the rotation of held wafer on the polishing pad [22]. The slurry used in the CMP process is dispensed through a slurry arm with the help of polishing pad conditioner, and the polishing pad surface is refreshed for each polishing process so that global planarization and polishing can be achieved [22]. During CMP process, the wafer is polished through abrasive and chemistry, and the complicate interaction between pad asperity, slurry, and wafer surface is described in **Figure 5** in a microscale observation [22]. For CMP process, readers are directed to Refs. [21, 22] for more details.

2.2 Frontside-release process for flexible silicon membrane

Frontside-release process utilizes SOI wafers and, in general, consists of active device fabrication, frontside-release hole, or structure patterning, releasing protection coating and release etching. Two release etching strategies are usually employed: one way is to remove BOX layer in SOI and fully release the device layer. RIE or DRIE is used to etch the Si device layer depending on the required etching depth and expose the BOX layer to HF etchant for releasing.

The other approach is to remove bulk silicon carrier in SOI and fully release the structures consisting of both device and BOX layers. Therefore, Si isotropically etching is required for releasing, and XeF_2 is mostly employed. Pure SF_6 plasma can also etch silicon isotropically.

Different from CMOS process, the thickness of FD-SOI device layer for MOSFET is approximately in the order of 100 nm, and the thickness of SOI device layer for flexible electronics can be as thick as 10 μm in many applications. Therefore DRIE is essential to form high aspect ratio trenches for exposure of BOX or silicon substrate to etchant. Following DRIE, the release can be achieved by either removing BOX or undercutting silicon substrate from frontside. HF etchant is used to remove BOX

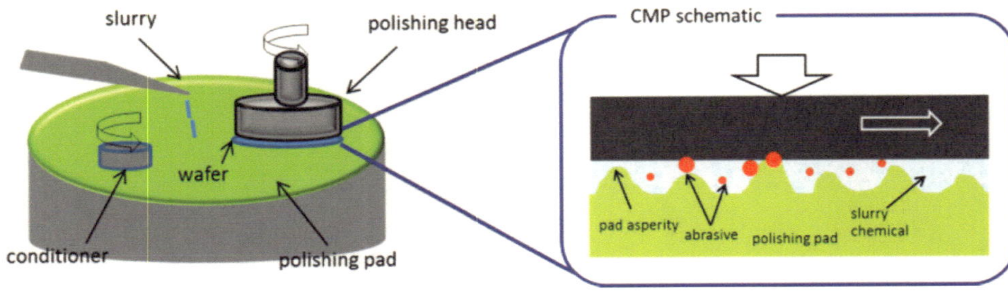

Figure 5.
The schema of conventional CMP process [22]. Copyright (2018) with permission from IntechOpen.

and XeF_2 is used to undercut silicon. Moreover, for release by undercutting silicon below BOX layer, a protective layer is necessary to protect other silicon parts from etching. Deposition of the protective layer onto the sidewall of trenches must be performed. Compared to PVD and CVD, ALD method can deposit high conformal and continuous protective layer inside the trench. The protective layer can be made of silicon oxide or alumina. In the following sections, we first introduce DRIE and XeF_2 RIE processes, and then we will describe how these two technique are utilized in the frontside-release processes.

2.2.1 Deep reactive-ion etching and XeF_2 RIE

2.2.1.1 Deep reactive-ion etching

DRIE is an extension of the traditional RIE process and is a highly anisotropic etch process. Different from traditional RIE, DRIE can be used to create vertical (90°) etch profiles, deep penetration, and holes with high aspect ratios. So far it has been used to fabricate capacitors for deep trench DRAM, TSVs, and microphotonic structures. With the help of novel thermal budget and by-product redeposition management, DRIE can pattern more than 5 μm silicon or even thru the wafer with cycling of two processes:

1. Plasma-induced deposition of a polymeric layer as passivation layer using C_4F_8 as working gas [18].

2. Anisotropic removal of passivation layer on the bottom followed by an isotropic Si chemical etch, and SF_6 is usually employed as working gas for etching of Si [18].

The DRIE process is shown in **Figure 6** and consists of six steps. At first, the polymeric passivation layer is coated overall to protect the sidewalls from chemical attack in the etching step (**Figure 6a**). Following that is the etching step; first the passivation layer on vertical surface (trench bottom) is removed through electrical field-accelerated ions (**Figure 6b**), and after the removal of passivation layer on trench bottom, the trench bottom is isotropically etched (**Figure 6c**). This isotropic etching usually lasts a few seconds, and the working gas mostly is a fluorine-based gas such as SF_6. Followed by the etching step, a deposition step is performed for a few seconds to coat the overall polymeric passivation layer (**Figure 6d**), which is similar to the first step. Then the etching step is repeated (**Figure 6e**), which is similar to step 2 and step 3 [18]. The removal of passivation layer on vertical surfaces is much faster than on horizontal, since the ions are accelerated in vertical direction. After the removal of passivation layer on the trench bottom, the further etchants start etching the trench bottom, and simultaneously polymeric passivation layer of the sidewall slows the lateral etch rate [18]. To achieve the desired depth of TSVs, these etching and deposition steps are repeated several times (**Figure 6f**) [18]. The DRIE process involves six steps, and the performance of each step is controlled by a significant number of parameters such as gas flows, the power of the inductively coupled plasma or the platen source, time, etc. [18].

There are quite a few parameters that can significantly influence the DRIE process profile, such as gas flows, the power of the inductively coupled plasma or the platen source, time, etc. For 1 μm line and hole, scallops were deeper in the top (~40 nm) and none in the middle (<5 nm) and minimal (~20 nm) in the bottom for both holes and lines for the optimized recipe [23]. **Figure 7** shows how the process parameters influence the DRIE process properties. Scallops in top and in the bottom

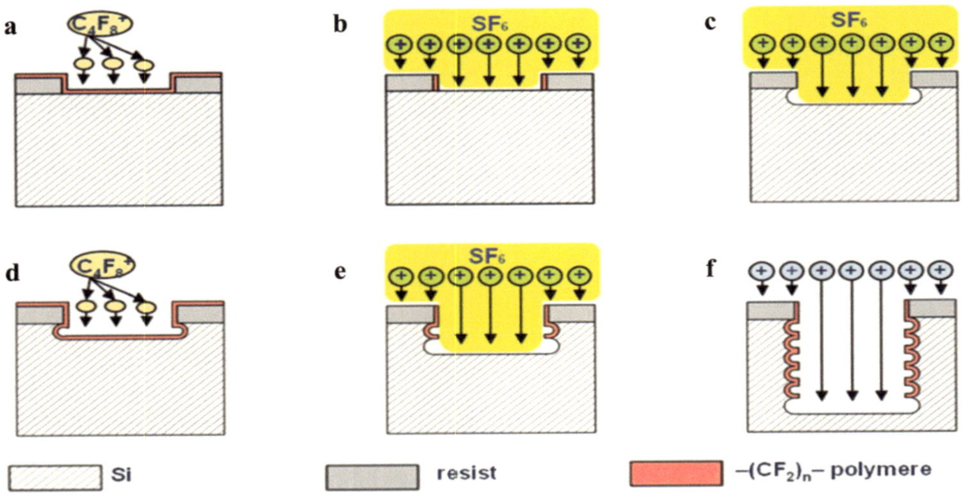

Figure 6.
Bosch process scheme. (a) Deposition of a conformal C_4F_8 passivation layer, (b) directed removal of the passivation layer by ions, (c) isotropic etching with SF6, (d) deposition of a conformal C_4F_8 passivation layer, (e) passivation removal and isotropic etching, and (f) alternating steps (b)–(e). reproduced with permission [18]. Copyright 2016, springer international publishing Switzerland.

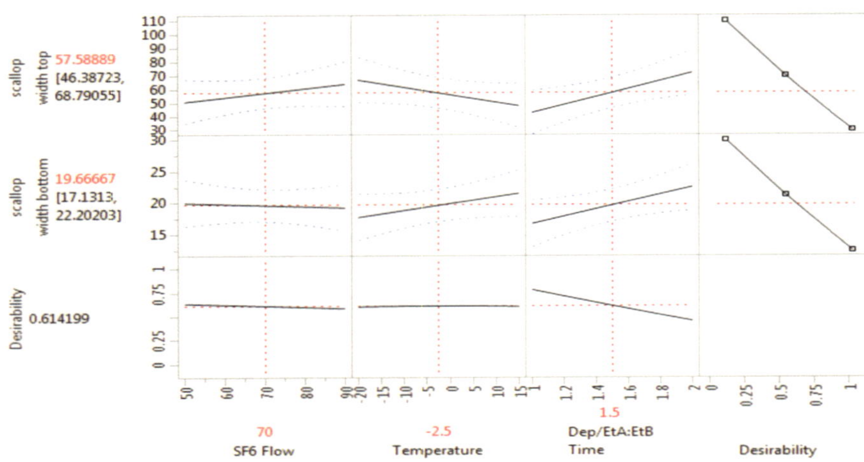

Figure 7.
DRIE profile scalloping prediction and desired profiles [23].

showed that a lower etch time results in less scalloping. Those etches are isotropic, so lowering the time lowers the etch distance in all directions. It also appears to be a weak but somewhat significant evidence for dependence on temperature. However, these trends are in the opposite directions, so there is no optimal temperature for the minimal scalloping. SF_6 flow shows no measured statistical significance to the scalloping or undercut.

2.2.1.2 XeF_2 RIE

Xenon difluoride (XeF_2), bromine trifluoride (BrF_3), chlorine trifluoride (ClF_3), and fluorine (F_2) are widely used to etch silicon [24]. Compared to other silicon etchants, XeF_2 has unique advantages like gas-phase isotropic etching, high selectivity for silicon, and ease of operation [24]. At room temperature and atmospheric pressure, XeF_2 is white and in solid state [24]. However, when XeF_2 is at a pressure smaller than 4 torr, the XeF_2 solid will transform into a gas state [24]. Since the gas etching process is simple to operate, XeF_2 etching process is widely performed by using the pulse etching system [24]. The XeF_2 pulse etching process can be controlled by process parameters such as XeF_2 pressure, etching time for a single cycle, and the number of etch cycles [24]. **Figure 8** shows the micromachining mechanism of XeF_2 etching.

2.2.2 Release through removal of BOX

In this release process, the SOI BOX layer is patterned and exposed using RIE or DRIE, followed by HF wet etching and critical point dry to fully release the structures above the BOX layer.

This release process involves three steps. In the first step, trenches are formed through RIE or DRIE to expose BOX (**Figure 9b**). When the silicon layer is thicker than 10 μm and aspect ratio is more than 10:1, DRIE is essential to expose BOX. For thin SOI, the exposure can also be performed by RIE. The second step is the deposition of a protective layer (**Figure 9b**). The Protective layer can protect other parts of silicon oxide from damage by etching, and the materials such as Si_3N_4 and PR can be used as a protective layer in this process. In the last step, the wafer is immersed in a HF-contained solution, which removes the exposed BOX. **Figure 9c** shows that the BOX is already partially removed, and **Figure 9d** shows that the BOX is fully removed through HF-contained solution. After the release process, the wafer will be transferred to a flexible substrate, and a bonding process will be performed to bond wafer to flexible substrate.

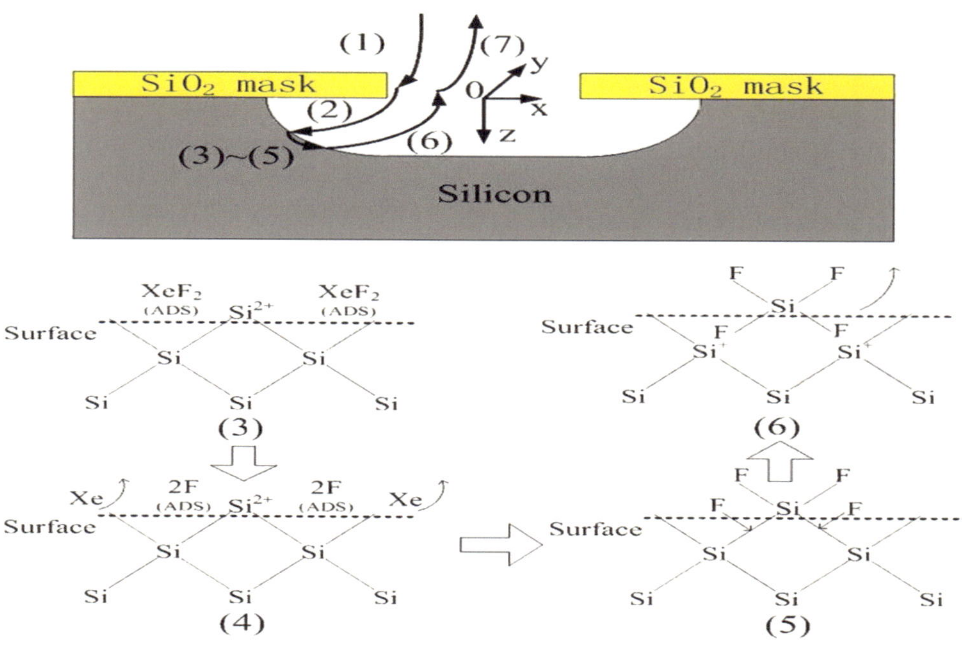

Figure 8.
Schema of interaction between XeF_2 and Si by using XeF_2 RIE to etch Si. (1) XeF_2 gas diffused from the reactor to the external surface of the etching window. (2) XeF_2 gas diffused from the etching window through the etched Si cavity to the silicon surface. (3) adsorption of XeF_2 at the silicon surface. (4) dissociation of XeF_2 molecule into fluorine atoms (F) and xenon (Xe) gas. (5) formation of Si-F bond and adsorption of SiF_4 at the silicon surface. (6) SiF_4 at the external surface is desorpted from Si surface. (7) the products are transferred from the wafer surface to the reactor. Reproduced with permission [24]. Copyright 2012, IEEE.

Zhou et al. utilized this release process to release their strained nanomembrane. In their paper for fast flexible electronics with strained silicon nanomembrane, the strips are released in a 4:1 diluted HF (49% HF) solution in which the BOX layer is selectively etched away [25]. **Figure 10** shows the process for release of silicon nanomembrane from Si handling substrate.

2.2.3 Release through undercut of silicon

This release process is achieved by undercutting silicon substrate under the BOX through XeF_2 isotropic etching. DRIE is usually used to pattern top silicon device layer followed by protective coating and removal of BOX layer in RIE.

This process involves four steps, and we use **Figure 11** to describe this release process. At first, an oxide film such as PECVD SiO_2 is deposited atop the device as an etching buffer layer (**Figure 11a**). Following that, with the help of a PR mask, the exposed oxide layers are removed through RIE, and then the exposure of silicon

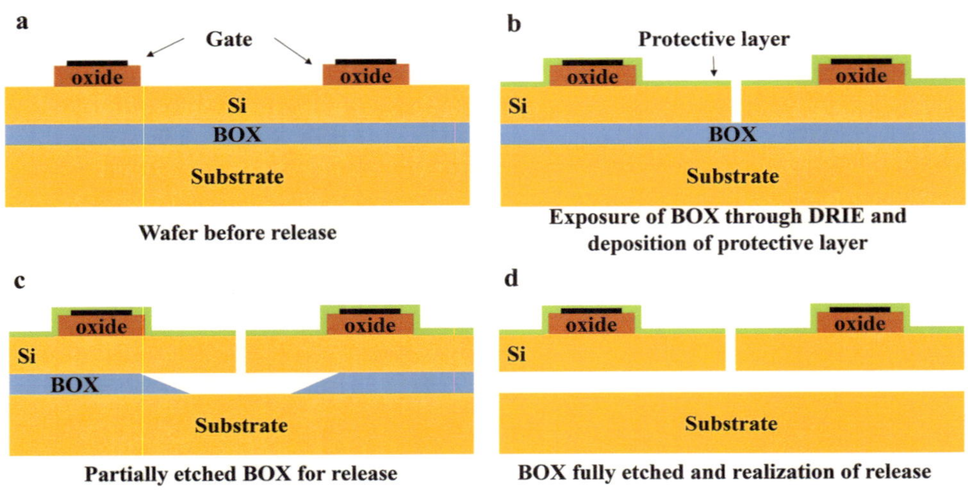

Figure 9.
The process flow for release through removal of BOX.

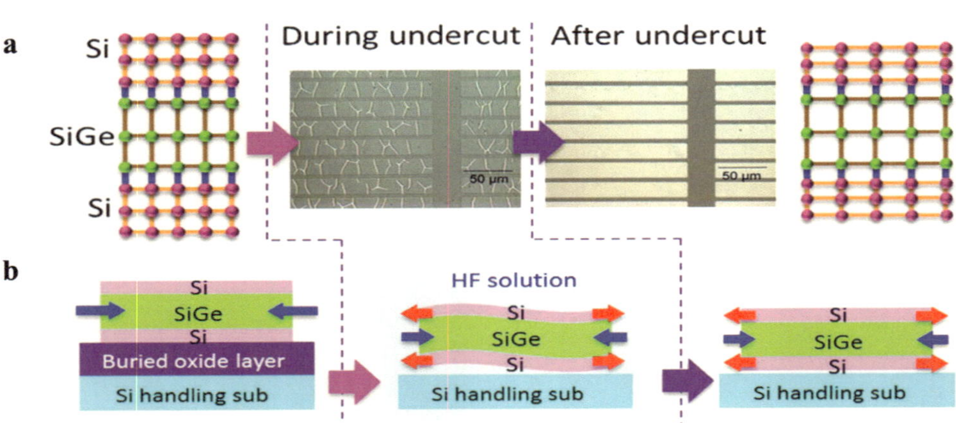

Figure 10.
(a) Atomic lattice schematic diagram showing the strain sharing principle. Optical images show the strained NM during release and after finishing release. (b) Process flow to implement the strain sharing principle and the release. Reproduced with permission [25]. Copyright 2013, nature publishing group.

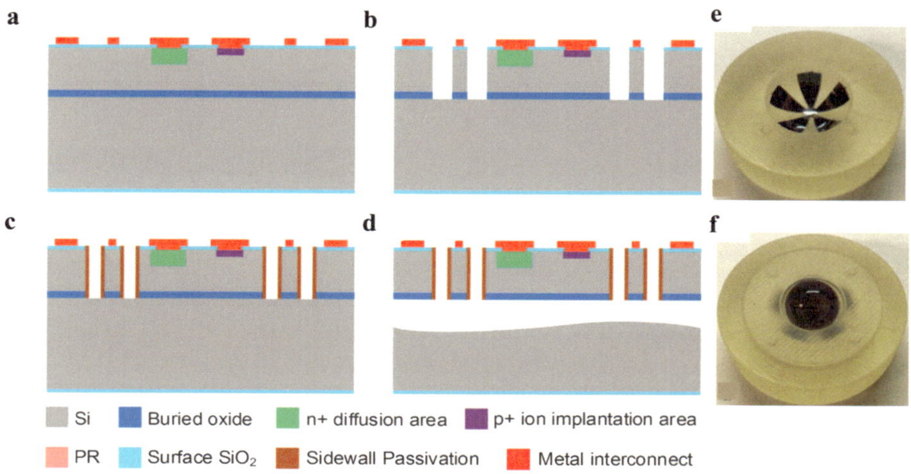

Figure 11.
DRIE process flow and fabricated monocentric imager: (a) after fabrication of photodiode circuitry, (b) pattern tessellated structures thru Si device and buried oxide layers, (c) sidewall passivation, (d) released device by XeF_2 etching, (e) a released and curved device transferred into a hemispherical fixture, and (f) a mounted monocentric imager. Reproduced with permission [35]. Copyright 2016, nature publishing group.

under the BOX is performed by using DRIE or RIE (**Figure 11b**). Following exposure septs, protective layer is coated overall to protect other parts of silicon from damage through etching (**Figure 11c**), and the materials such as Al, PR, GaN, and SiO_2 can be used as protective layer in this process. After that, a RIE etching is performed to remove the protective layer at the bottom of the trenches. At last, the wafer is placed in XeF_2 RIE to etch the silicon under the BOX. Once the undercuts meet with each other, the SOI and the BOX is completely released from the bulk substrate (**Figure 11d**). After the release process, the wafer will be transferred to flexible substrate, and a bonding process will be performed to bond wafer to flexible substrate.

Wu et al. (2016) employed this release process to fabricate a silicon-based flexible imager [26–33], and **Figure 11** shows the process flow to release an imager from the carrier substrate and the fabricated mounted monocentric imager. Sevilla et al. used this release process to fabricate a silicon-based flexible FinFET. **Figure 12** shows the basic steps to release FinFET from the carrier substrate, the fabricated FinFET, and the flexible FinFET wafer [34].

2.3 Backside-release process for flexible silicon membrane

Instead of etching the BOX and undercutting silicon substrate under the BOX, backside-release process etches the silicon substrate from the backside. Lapping, CMP, and RIE can be employed in this release process. The mechanisms of CMP and XeF_2 RIE are already introduced in this chapter, and in this section the comparison between CMP, XeF_2 RIE, and lapping is described. Following that, the process flow is described through the example in which lapping, CMP, and RIE are employed to thin the silicon substrate.

2.3.1 Comparison between CMP, RIE, and lapping

CMP, RIE, and lapping are used to realize backside-release, and we compare these tree fabrication processes to know how to choose suitable fabrication process.

The mechanisms of CMP and XeF_2 RIE are already introduced in previous sections, and lapping is a mechanical process in which a pad is used with polishing liquid to remove excess silicon from a wafer substrate. Lapping takes place between

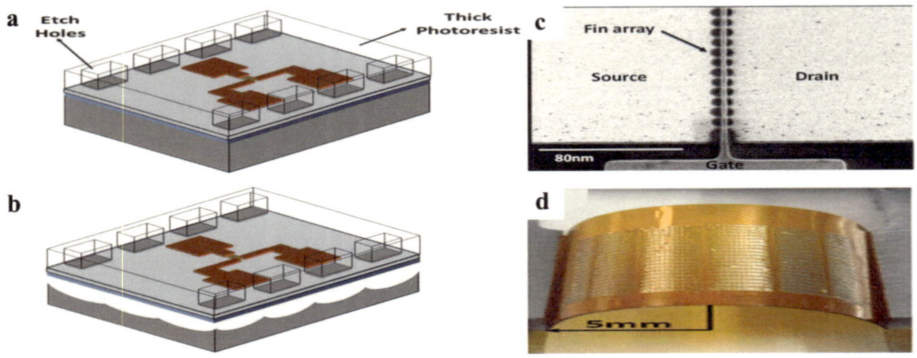

Figure 12.
(a) Spin coat of thick (7 μm) photoresist and hole patterning, (b) cavern formation beneath BOX due to XeF_2 etchant, (c) top view of fins after the gate etch process, which is a complex task performed with a combination of reactive-ion etching and wet cleans, and (d) FinFET silicon fabric at minimum device scale bending radius (5 mm). Reproduced with permission [34]. Copyright 2014, John Wiley & Sons, Inc.

two counter-rotating cast iron plates and either an abrasive film or slurry. To adjust the penetration of the film/slurry, the wafers either spin faster or experience a heavier load to fit the target specification.

The surface roughness of silicon is about 50 nm by using CMP, and through lapping the surface roughness of silicon can achieve 1 μm. The surface roughness of silicon by using RIE is worst about 10 μm. The cost of RIE is the most expensive, because this process needs also a working gas and vacuum environment. The CMP process is more expensive compared to lapping, because CMP consumes chemicals while lapping involves mechanical polish only. For thinning substrate, man can chose suitable process depending on the surface roughness and the cost.

Figure 13 shows three present backside thinning processes by using CMP, RIE, or lapping. For thinning through lapping and XeF_2 RIE (**Figure 13a**), at first the substrate is reduced to exact thickness (usually about 50 μm) by lapping for cost saving, and after that the resulting surface micro-crack damages induced during the lapping process are removed by XeF_2 etching processes with the buried oxide layer as the etch stop layer. Thinning process through anisotropic RIE (**Figure 13b**) possesses advantage high etch rate about 20 μm/min and disadvantage high surface roughness. At first the wafer is turned upside down, and then the substrate is thinned through RIE. The thickness of substrate is controlled by a measurement. Besides RIE and lapping, backside-release can be also performed through CMP, and **Figure 13c** shows this thinning process. Compared with RIE and lapping, the etch rate of CMP is much lower about 0.5 μm/min, and the surface roughness is best at about 50 nm. Usually for cost saving, before CMP process, the substrate can be thinned through lapping, and after lapping process expensive and more precise CMP is performed to further thin the substrate.

2.3.2 Process flow of backside-release process

2.3.2.1 Backside-release process using lapping and XeF_2 RIE

Lapping and XeF_2 etching can thin the SOI wafer from the backside all the way to the BOX layer with a clean surface finish due to high selectivity between SiO_2 and Si in XeF_2 RIE. This process involves three steps. At first the wafer is coated by protective layer to protect the parts of wafer that shall not be etched from damage through isotropic XeF_2 RIE. Following that, the substrate can be thinned to certain thickness by lapping. At last, the rest of Si substrate is removed by XeF_2 RIE, and

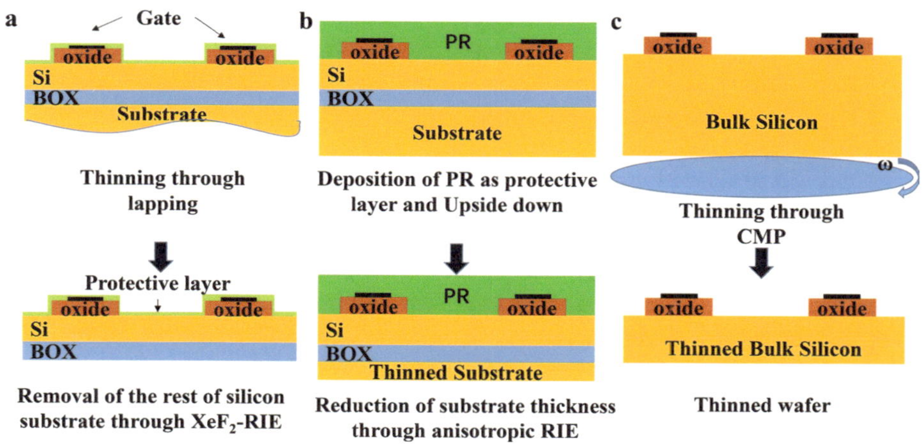

Figure 13.
The schema for three present backside thinning processes by using CMP, RIE, or lapping. (a) Thinning through lapping (first step) and XeF_2 RIE (second step), the BOX serves as stop layer for XeF_2 RIE, (b) thinning through RIE by using working gas such as CF_4 and SF_6, and (c) thinning through lapping (first step) and CMP (second step).

the BOX serves as stop layer for XeF_2 RIE. With the help of XeF_2 RIE and stop layer, the resulting surface micro-crack damages induced during the lapping process can be removed. After thinning process, the wafer is transferred to a flexible substrate, and a bonding process will be performed to bond wafer to flexible substrate.

Hsieh et al. used this process to fabricate a biocompatible flexible IC. At first a wafer lapping machine is used to thin the Si wafer substrate, and the thickness is reduced to ~50 μm [28]. Following lapping, a dry XeF_2 etching process with BOX as etch stop layer is performed to remove the surface micro-crack damages that are caused by the lapping process [28]. Liu et al. employed this process to fabricate a spherical flexible CMOS retina chip. They thinned the backside Si to a thickness of around 50 μm by mechanical lapping, and after that a dry etching such as XeF_2 or RIE is used to etch the Si substrate down to around 10μm thickness and remove the surface micro-crack damages induced during the lapping process [33].

2.3.2.2 Backside-release process using RIE

Instead of lapping and XeF_2, RIE can be directly used to thin the backside bulk silicon substrate, although it results in high roughness.

This backside-release process consists of deposition of photoresist to protect wafer during etching and RIE etching with thickness measurement of the substrate. In this process, the substrate can be traditional bulk substrate or SOI substrate, because the thickness is controlled by a measurement instead of a stop layer. Moreover, the protective layer against etchant is not needed for this process, because an anisotropic RIE is used to thin substrate. The PR is coated to protect the ultrathin wafer from mechanical damage such as scrape and fracture. Working gases such as CF_4 and SF_6 are used to etch the silicon in RIE.

Sevilla et al. have used this approach to fabricate a flexible nanoscale high performance FinFET [36]. **Figure 14a** shows wafer with FinFET before the release process. First step for this process is deposition of PR that servers as protect layer against mechanical damage (**Figure 14b**). After deposition of PR, the wafer is turned upside down, and the substrate is thinned through RIE (**Figure 14c**). The thickness of substrate is controlled by measurement; when the thickness is the same to the plan, the thinning is finished. Otherwise, the wafer is placed in RIE again for further reduction of substrate. **Figure 14d** shows the thinned wafer, and at last the PR layer is removed (**Figure 14e**). If the SOI is not very thin and the surface is hard, this PR layer is not anymore necessary for this release process.

2.3.2.3 Backside-release process using CMP

Besides RIE and lapping, backside-release can be also performed through pure CMP process. The etch rate of CMP is much lower than RIE and lapping, but the surface roughness is the best and in the order of 10 nm or less. Since CMP process is very slow, it usually starts with a thin substrate, for example, 100–200 μm or after a lapping process with reduced thickness for cost saving purpose.

Dumas et al. use CMP to fabricate curved focal plane detector array for wide field cameras. To spherically curve the device, they used CMP to thin the substrate. In their experiment, the process is designed to obtain a component thickness of 50 μm [37]. They have demonstrated that 10 × 10 mm^2 silicon samples thinned down to 50 μm could be curved in concave and convex shapes, down to a bending radius of 40 mm [37]. The curved detector is showed in **Figure 15**.

2.4 Transfer processes and bond technique

After the release, the released membranes are transferred to flexible substrate and then bonded to flexible substrate. Now we introduce transfer process and bond technique for silicon-based flexible electronics.

2.4.1 Transfer through polymer stamp

In this transfer process, a photoresist or similar polymer layer is deposited, and then a flat piece of polymer such as poly(dimethylsiloxane) PDMS serves as stamp, which conformally contacts the top surface of the wafer. When the stamp is in contact with the PR, it is carefully peeled up with the released thin membrane. The interface between stamp and photoresist must be strongly bonded, and the wafer is transferred on a flexible substrate. The flexible substrate can be polyimide substrate or liquid crystal polymer (LCP) substrate, and the polyimide adhesion

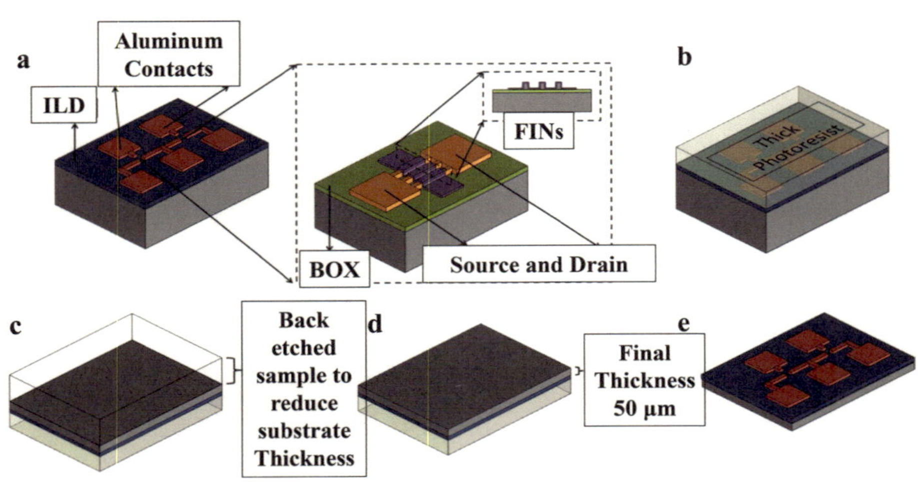

Figure 14.
Process flow for the fabrication of flexible FinFET: (a) produced FinFET devices on SOI substrate (90 nm SOI with 150 nm BOX); (b) deposition of PR to protect chip from damages induced by back etch process; (c) FinFET devices etched from backside using RIE process; (d) Si substrate thinned to 50 μm; and (e) removal of PR. Reproduced with permission [36]. Copyright 2014, American Chemical Society.

promoter is spin coated on substrate; once the wafer is brought to polyimide, the wafer is baked to cure polyimide adhesion promoter. At last the PR and stamp are striped. This transfer process is the same to the process shown in **Figure 16a**.

Figure 16a shows a transfer process developed by Menard et al. In this process, bendable single crystal silicon thin film transistors are printed on plastic substrates. At first they brought a flat piece of PDMS that served as stamp into conformal contact

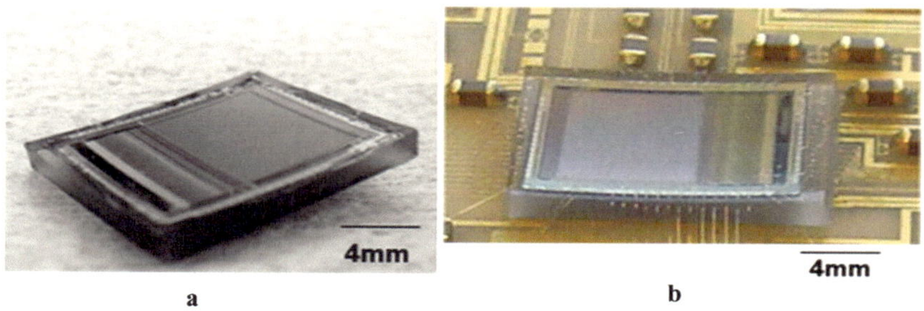

Figure 15.
(Color online) Pictures of curved microbolometer. (a) The thinned curved component on a glass holder. (b) This curved bolometer is bonded onto an electrical board. Reproduced with permission [31]. Copyright 2012, The Optical Society.

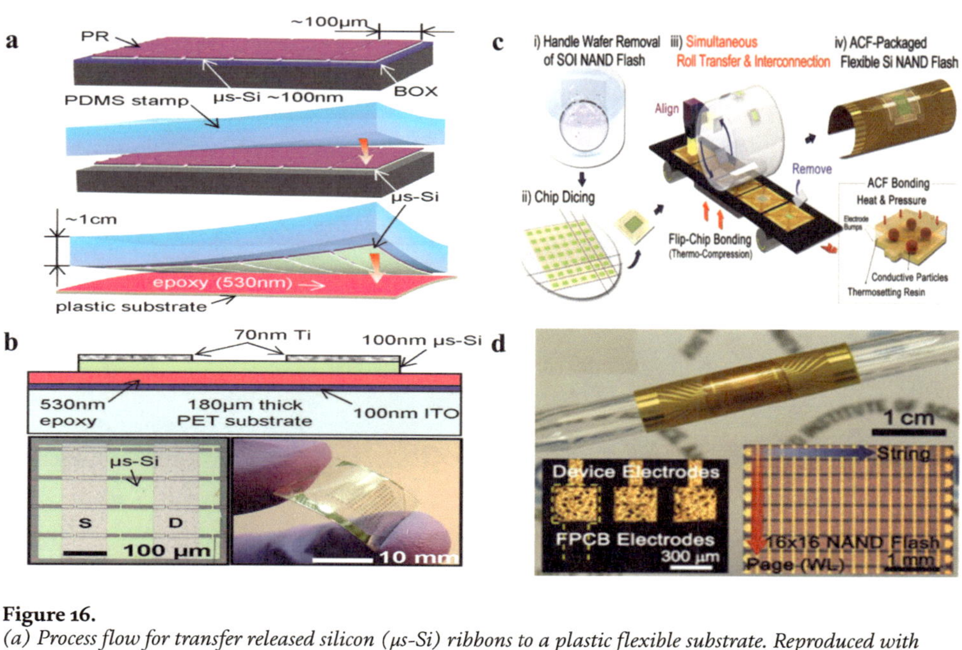

Figure 16.
(a) Process flow for transfer released silicon (μs-Si) ribbons to a plastic flexible substrate. Reproduced with permission [38]. (b) the schematic structure of a flexible thin film transistor with high performance, which was transferred on a PET substrate. The bottom insets show optical images of a device array. In insets, we can see that each device consists of four interconnected microstrips of μs-Si (100 nm thick). Reproduced with permission [38]. Copyright 2005, American Institute of Physics. (c) Process flow for the fabrication of the ACF-packaged flexible Si NAND flash memory using roll-based thermo-compression bonding. (i) Fabricated NAND flash on an SOI wafer was bonded with transfer glass. Release processes are realized by wet etching of the BOX. (ii) NAND chip bonded on transfer glass was separated by dicing. (iii) the released memory was transferred and interconnected on FPCB by using roll-based transfer. (iv) fabricated ACF-packaged Si f-NAND. As shown in inset, the adhesion and interconnection of the electrode bumps of the device and the FPCB are realized by thermosetting resin and conductive particles, respectively. Reproduced with permission [39]. Copyright 2016, John Wiley & Sons, Inc. (d) Photograph of the highly compliant ACF-packaged f-NAND wrapped on a glass rod (diameter of 7 mm). The OM image of the electrode area (left) and the active device area (right) are shown in insets. Reproduced with permission [39]. Copyright 2016, John Wiley & Sons, Inc.

with PR layer on the surface of the wafer and then carefully peeled back to pick up the released wafer with silicon (μs-Si) ribbons [38]. The interaction between the PR and the PDMS must be strong enough to bond them together for removal, with good efficiency [38]. A 180 μm thick polyethylene terephthalate (PET) plastic sheet coated with a 100 nm thick indium tin oxide (ITO) was used as the flexible substrate [38]. A dielectric layer of epoxy was used to enhance the adhesion between released wafer and flexible substrate and was spin coated on flexible substrate [38]. Bringing the PDMS with the μs-Si on its surface into contact with the warm epoxy layer and then peeling back the PDMS led to the transfer of the μs-Si to the epoxy [38].

Kim et al. demonstrate simultaneous roll transfer and interconnection of Si-based flexible NAND flash memory (f-NAND) based on highly productive roll-to-plate ACF packaging [39]. This process is described in **Figure 16b**.

2.4.2 Bonding SOI to flexible substrate

When thinned die or wafer is transferred on the flexible substrate, the bonding between die and flexible substrate must be performed in order to realize the electrical connection between die and other devices.

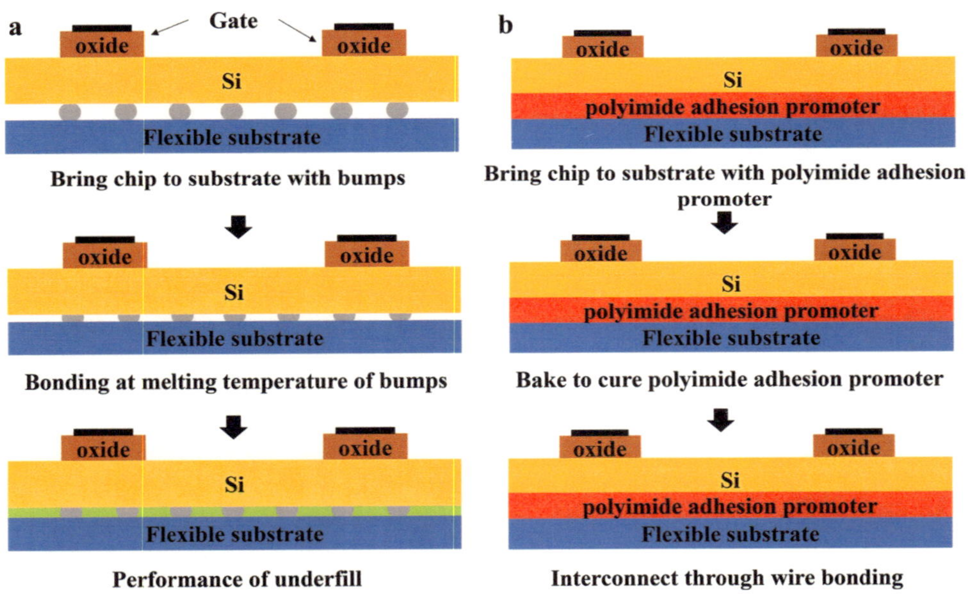

Figure 17.
(a) Process flow for flip-chip bond and (b) process flow for adhesion method.

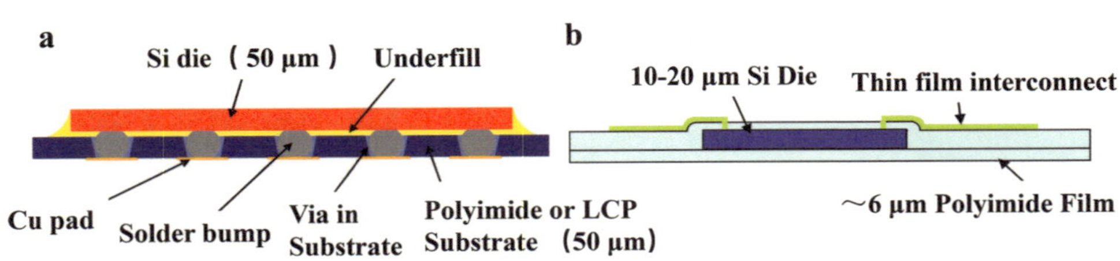

Figure 18.
(a) Illustration of polyimide or LCP substrate and solder assembly approach. Reproduced with permission [40]. Copyright 2008, IEEE. (b) Illustration of thinned Si die embedded in polyimide with thin film interconnect using adhesion method. Reproduced with permission [40]. Copyright 2008, IEEE.

Flip-chip bond and adhesion method can be used to bond released dies to flexible substrate. For the flip-chip bond, polyimide or liquid crystal polymer (LCP) can be used as flexible substrate. For adhesion method, the substrate is made of polyimide, and the interconnection of die is formed through wire bonding.

Flip-chip bond consists of four steps. First, the bumps and pads are fabricated on flexible substrate. Following that, the die is placed on the flexible substrate and aligned. Once the die and bumps are in contact, the bumps are heated at melting temperature, and then die is bonded to substrate. At last an underfill is performed. The materials such as SnPb and SnAg can be used as bumps.

For the adhesion method, a polyimide adhesion promoter such as a dielectric layer of epoxy is applied, so that the die can adhere to polyimide substrate. The polyimide adhesion promoter is spin coated on substrate, and once the wafer is brought to polyimide, the wafer is baked to cure polyimide adhesion promoter. At last the interconnection is formed through wire bonding (**Figure 17**).

Holland et al. [40] used flip-chip bonding to bond die to substrate. Different from our bonding process, they used immersion bump. **Figure 18a** shows thinned die flip-chip bonded on polyimide or LCP substrate. Moreover Holland et al. [40] used also adhesion method to bond die to substrate. Different from our adhesion method, they embed the thinned Si die in Polyimide (**Figure 18b**).

Menard et al. used a dielectric layer of epoxy as polyimide adhesion promoter that was spin coated on substrate to bond the die to polyimide substrate through adhesion method [38].

3. Conclusion

Theoretically, all devices such as transistor circuit, DRAM, NAND flash, and sensors that were fabricated through traditional Si-based CMOS process can also be fabricated in flexible forms by using appropriate release processes and transfer technique. We have mainly described two types of release processes: frontside- and backside-release. The frontside-release is realized by etching the BOX or undercutting silicon under the BOX in SOI wafer. The BOX layer etching is achieved in wet etching with HF-contained etchant, and the bulk silicon undercutting is achieved by XeF_2 isotropic etching. The backside-release process etches the Si substrates through CMP, lapping, or RIE. After releasing, the Si thin membrane with active devices is transferred to a flexible substrate. Polymer stamp transfer, flip-chip bond, or adhesion method can be used to bond released dies to a flexible substrate. By leveraging those silicon-based micromachining processes, flexible electronics can be achieved on top of current standard CMOS process and scale to large volume manufacturing.

Author details

Jiye Yang and Tao Wu*
School of Information Science and Technology, ShanghaiTech University, China

*Address all correspondence to: wutao@shanghaitech.edu.cn

References

[1] Zhang K et al. Fast flexible electronics using transferrable silicon nanomembranes. Journal of Physics D: Applied Physics. 2012;**45**(14):143001

[2] Muller R, Kamins T. Device Electronics for Integrated Circuits. New York: Wiley; 2003. p. 530

[3] Kwon JY et al. High efficiency thin upgraded metallurgical-grade silicon solar cells on flexible substrates. Nano Letters. 2012;**12**(10):5143-5147

[4] Weisse JM et al. Vertical transfer of uniform silicon nanowire arrays via crack formation. Nano Letters. 2011;**11**(3):1300-1305

[5] Que R et al. Silicon nanowires with permanent electrostatic charges for nanogenerators. Nano Letters. 2011;**11**(11):4870-4873

[6] Yu B et al. FinFET scaling to 10nm gate length. In: International Electron Devices Meeting; 2002; IEEE; 1998

[7] Boukai AI et al. Silicon nanowires as efficient thermoelectric materials. In: Materials for Sustainable Energy: A Collection of Peer-Reviewed Research and Review Articles from Nature Publishing Group. United Kingdom, London: World Scientific; 2011. pp. 116-119

[8] Sun B, Shao M, Lee S. Nanostructured silicon used for flexible and mobile electricity generation. Advanced Materials. 2016;**28**(47):10539-10547

[9] Kim T et al. Flexible, highly efficient all-polymer solar cells. Nature Communications. 2015;**6**:8547

[10] Yang L et al. Solution-processed flexible polymer solar cells with silver nanowire electrodes. ACS Applied Materials & Interfaces. 2011;**3**(10):4075-4084

[11] Cao Q , Rogers JA. Ultrathin films of single-walled carbon nanotubes for electronics and sensors: A review of fundamental and applied aspects. Advanced Materials. 2009;**21**(1):29-53

[12] Yamada T et al. A stretchable carbon nanotube strain sensor for human-motion detection. Nature Nanotechnology. 2011;**6**(5):296-301

[13] Moriceau H et al. New layer transfers obtained by the SmartCut process. Journal of Electronic Materials. 2003;**32**(8):829-835

[14] Benoist T et al. ESD robustness of FDSOI gated diode for ESD network design: Thin or thick BOX? 2010. IEEE International SOI Conference (SOI). 2010

[15] Pal A. MOS Fabrication Technology. New Delhi: Springer; 2015. pp. 19-42

[16] Rossnagel SM, Powell R, Ulman A. PVD for Microelectronics: Sputter Desposition to Semiconductor Manufacturing. Vol. 26. London: Elsevier; 1998

[17] Wanebo M et al. Molecular vapor deposition (MVD/spl trade/)—A new method of applying moisture barriers for packaging applications. In: Proceedings International Symposium on Advanced Packaging Materials: Processes, Properties and Interfaces; 2005; 2005

[18] Killge S, Neumann V, Bartha JW. Copper-based TSV: Interposer. In: 3D Stacked Chips. Switzerland: Springer; 2016. pp. 9-28

[19] Gogolides E et al. Si etching in high-density SF6 plasmas for microfabrication: Surface roughness formation. Microelectronic Engineering. 2004;**73**:312-318

[20] Vugts M et al. Si/XeF_2 etching: Reaction layer dynamics and

surface roughening. Journal of Vacuum Science & Technology, A: Vacuum, Surfaces, and Films. 1996;**14**(5):2780-2789

[21] Hong J et al. Removal rate and surface quality of the GLSI silicon substrate during the CMP process. Microelectronic Engineering. 2017;**168**:76-81

[22] Kim HJ. In: Rudawska A, editor. Abrasive for Chemical Mechanical Polishing, Abrasive Technology. Rijeka: IntechOpen; 2018

[23] Hamann ACAS. Smooth Sidewall Etching in the PT-DSE. 2014

[24] Xu D et al. Isotropic silicon etching with XeF_2 gas for wafer-level micromachining applications. Journal of Microelectromechanical Systems. 2012;**21**(6):1436-1444

[25] Zhou H et al. Fast flexible electronics with strained silicon nanomembranes. Scientific Reports. 2013;**3**:1291

[26] Dinyari R et al. Curving monolithic silicon for nonplanar focal plane array applications. Applied Physics Letters. 2008;**92**(9):091114

[27] Rim S-B et al. The optical advantages of curved focal plane arrays. Optics Express. 2008;**16**(7):4965-4971

[28] Hsieh C et al. A flexible mixed-signal/RF CMOS technology for implantable electronics applications. Journal of Micromechanics and Microengineering. 2010;**20**(4):045017

[29] Iwert O, Delabre B. The challenge of highly curved monolithic imaging detectors. In: High Energy, Optical, and Infrared Detectors for Astronomy IV. Germany: International Society for Optics and Photonics; 2010

[30] Kim DH et al. Stretchable, curvilinear electronics based on inorganic materials. Advanced Materials. 2010;**22**(19):2108-2124

[31] Dumas D et al. Infrared camera based on a curved retina. Optics Letters. 2012;**37**(4):653-655

[32] Blake T et al. Utilization of a curved local surface Array in a 3.5 m wide field of view telescope. Arlington; 2013, Defense Advanced Research Projects Agency Arlington Va Tactical Technology Office

[33] Liu C-Y et al. A contact-lens-shaped IC chip technology. Journal of Micromechanics and Microengineering. 2014;**24**(4):045025

[34] Sevilla GA et al. Flexible and transparent silicon-on-polymer based sub-20 nm non-planar 3D FinFET for brain-architecture inspired computation. Advanced Materials. 2014;**26**(18):2794-2799

[35] Wu T et al. Design and fabrication of silicon-tessellated structures for monocentric imagers. Microsystems & Nanoengineering. 2016;**2**(1)

[36] Torres Sevilla GA et al. Flexible nanoscale high-performance FinFETs. ACS Nano. 2014;**8**(10):9850-9856

[37] Dumas D et al. Curved focal plane detector array for wide field cameras. Applied Optics. 2012;**51**(22):5419-5424

[38] Menard E, Nuzzo RG, Rogers JA. Bendable single crystal silicon thin film transistors formed by printing on plastic substrates. Applied Physics Letters. 2005;**86**(9):093507

[39] Kim DH et al. Simultaneous roll transfer and interconnection of flexible silicon NAND flash memory. Advanced Materials. 2016;**28**(38):8371-8378

[40] Holland B et al. Ultra-Thin, Flexible Electronics. Piscataway, New Jersey, US: IEEE; 2008. pp. 1110-1116

Chapter 4

CMOS Compatible Wet Bulk Micromachining for MEMS Applications

S. Santosh Kumar and Ravindra Mukhiya

Abstract

Wet bulk micromachining of silicon is a convenient and economical method for realizing various silicon-based microsensors and actuators. Tetramethylammonium hydroxide (TMAH) based anisotropic wet etching is popular due to it being less toxic and CMOS compatible. The etch rate of TMAH depends on the wafer's crystal plane orientation and temperature/concentration of solution. While using TMAH to realize a pressure sensor diaphragm, the etching of {111} planes causes underetching, causing a deviation in the intended size of the diaphragm, inducing variation in the designed characteristics of the device. It is necessary to estimate and minimize these deviations. Experiments were designed and the rate of etching for (100) and (111) planes using 25 wt.% TMAH have been determined at different temperatures. Linear fit equations are obtained from experimental data to relate the underetch per unit depth to the solution temperature. These findings are extremely useful in the fabrication of silicon diaphragms with precise dimensions. While using anisotropic wet etchants to realize proof mass for accelerometers, the etchants attack the convex corners. This necessitates a suitable design of compensating structure while realizing microstructures with sharp convex corners. Experimental studies are carried out to protect convex corners from undercutting and the results are reported.

Keywords: TMAH etching, piezoresistive pressure sensor, diaphragm, underetching, convex corner compensation

1. Introduction

The process of bulk micromachining is carried out in order to etch out a significant portion of silicon from silicon substrate resulting in structures created out of silicon for different applications. Some of these structures include: diaphragms (for pressure sensors) [1], proof-mass (for accelerometers) [2], nozzle (for inkjet printer) [3], random pyramids on Si surface (for solar cells) [4], and cantilevers [5]. Aqueous TMAH is often used for realization of structures in silicon using wet anisotropic bulk micromachining. It offers an attractive low-cost alternative to dry bulk micromachining technique, which uses gases and expensive equipment. TMAH is favoured choice over other alternatives like KOH, EDP etc. for wet anisotropic etching due to complementary metal-oxide-semiconductor (CMOS) compatibility and less toxicity.

IntechOpen

In industrial microelectronics process line or CMOS-MEMS processes, it is preferable to replace KOH with TMAH, to avoid contamination due to Potassium (K+) ions. However, owing to a high undercutting ratio of TMAH, it entails a dedicated study on compensation structures. This has inspired us to carry out experimental studies using TMAH.

TMAH etching rate is generally governed by: the orientation of the Si wafer, the temperature of aq. TMAH solution and the TMAH concentration in aqueous solution [6]. Although TMAH etches aluminium when it comes in contact with it, it is possible to carry out post CMOS bulk micromachining with aluminium metallization using TMAH by dissolving silicon and oxidizers like ammonium peroxodisulphate into the solution [7, 8].

The undercutting of convex corners and non (111) planes in Si is helpful in realization of freely suspended microstructures using wet anisotropic etchants [9, 10]. However, when mesa type structures having convex corners are desired, these effect of undercutting of convex corners have to be checked and avoided, if possible. It has been observed that TMAH shows a high undercutting ratio in comparison to KOH. In this chapter, experimental studies and analysis to protect convex corners and non {111} crystal planes from undercutting during TMAH etching is described.

In this chapter, we present the studies related to realization of diaphragm (for piezoresistive pressure sensor) and proof-mass (for accelerometer) using aq. 25 wt. % TMAH solution. The importance of precise etching of diaphragm in piezoresistive pressure sensor is discussed in Section 2. Experimental evaluation of TMAH etching is carried out to determine the etch rate in (100) and (111) planes of silicon at different temperatures and is discussed in Section 3. The underetching of silicon diaphragm during TMAH etching is also determined to design proper dimensions of openings in mask sets. Corner compensation structures for TMAH, experimental details, and design analysis and discussion are presented in Sections 4–6, respectively.

2. Piezoresistive pressure sensor

Usually, a piezoresistive pressure sensor has four piezoresistors on a silicon diaphragm, close to the edges of the diaphragm, connected in a Wheatstone configuration. When the diaphragm is subjected to a pressure load, the deflection of the diaphragm leads to stress generation on the diaphragm. As the piezoresistors are placed near the surface of the diaphragm, they experience these stresses, leading to a change in the resistance of the piezoresistors. Prior to fabricating the sensor, the placement of piezoresistors on the diaphragm is optimized using finite element method (FEM) based tools.

Diaphragm in (100) silicon can be fabricated using TMAH by exploiting its anisotropic etching property. TMAH etches different silicon planes with different etch rates but it has a high etch rate in <100> direction. Subsequent to the groundbreaking paper by Tabata et al. [11], much research has been dedicated to analyse the etch rate of silicon for different crystallographic directions [6, 12, 13]. While realizing diaphragms using aq. TMAH etching in (100) silicon wafers, the expected position of the piezoresistors (determined using FEM based design simulations) may shift due to etching in {111} planes. This leads to the phenomenon of underetching, where the TMAH etches under the etching mask in a (100) wafer. Underetching may cause the piezoresistors to be shifted from their pre-planned position on the diaphragm, as shown in **Figure 1**. This induces variation in the designed characteristics of the device. Therefore, it is necessary to estimate and

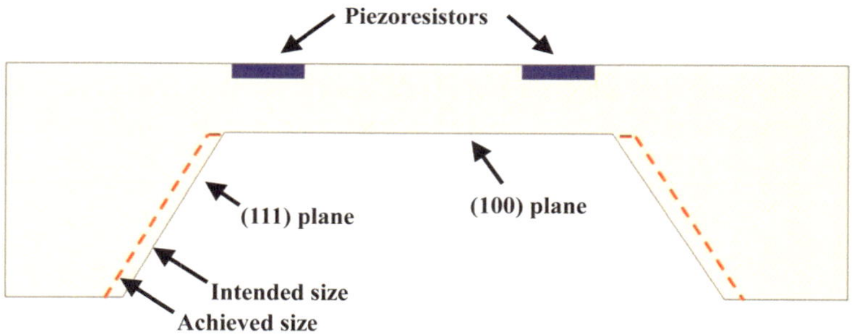

Figure 1.
Cross-section of diaphragm of pressure sensor showing the displacement of position of piezoresistors due to etching by TMAH in (111) planes.

minimize the deviation, especially in square diaphragms, where precise placement of piezoresistors is of prime importance for optimum performance of the device.

3. Experimental evaluation of TMAH etching for pressure sensor diaphragm

For the purpose of this study, experiments are designed and the etch rates in the {100} and {111} planes for 25 wt.% TMAH have been determined at different temperatures. Through these experiments, the underetching per unit depth of etching is also determined. This data can be used for selection of proper dimensions (based on the selected temperature of etching during fabrication) in the mask sets, compensating for the change in dimensions caused due to underetching. This method will ensure that diaphragms with accurate dimension are realized after etching.

The cross-section of the etch profile obtained during etching, while fabricating diaphragm structures on (100) Si substrate using TMAH is shown in **Figure 2**. The figure distinctly depicts anisotropic etching by TMAH and also shows the underetching below the mask due to etching in {111} planes. Alternatively, we can

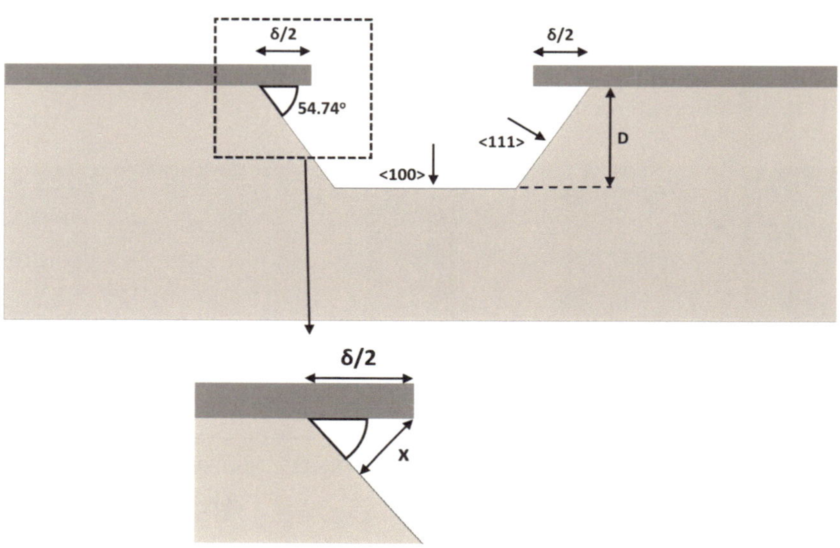

Figure 2.
TMAH etch profile depicting etching in (100) and (111) planes.

also say that underetching is dependent on etch rate of TMAH in <110> direction. If we consider etching for time t, the <100> direction etch rate (ER) can be calculated by determining the etch depth (D) and dividing it by t. The <111> direction etch rate (x/t) can be calculated by determining the underetching (δ/2) in the mask and subsequently applying the formulas as shown in Eqs. (1)–(3) [14].

$$\frac{x}{\delta/2} = \sin 54.74 = \frac{\sqrt{2}}{\sqrt{3}}, \delta = \sqrt{6}x \tag{1}$$

$$\text{ER of } \{111\} \text{ planes} : \frac{x}{t} = \frac{\delta}{\sqrt{6}t}, t = \text{etch time} \tag{2}$$

$$\text{Underetch per unit depth} : \frac{\delta/2}{D} \tag{3}$$

To ascertain the nature of etching in 25 wt.% aq. TMAH for (100) silicon at different temperatures, (100) Si n-type wafers with resistivity values in the range of 5–7 Ω-cm and a thickness of 527 ± 2 μm are selected. Subsequently, a thermal SiO_2 layer of 1 μm thickness is grown (and afterward patterned) for use as masking layer. A single level mask is used to obtain the required diaphragms. The diaphragms etched using TMAH are than analyzed for feature dimensions to determine the (111) and (100) planes etch rates. Each sample was etched in TMAH at different temperature (with an accuracy of ±1°C). For maintaining a constant concentration of TMAH in the aqueous solution, during the course of the experiment, suitable amount of deionized (DI) water was added after each experiment [15]. The etching on each sample was performed for 5 h. Six samples were used for six different temperatures.

The depth of the cavity (behind the diaphragm) for different samples is determined using a surface contact profilometer. The measured value is used to obtain the <100> direction etch rate. To obtain the <111> direction etch rate for different samples, the initial and final sizes of the square alignment mark are compared. The initial size of alignment mark would be equal to the size of cavity opening in the mask set. This determines the underetching by determining the length of hanging oxide. These measurements are carried out using scanning electron microscope (SEM). The underetch (δ/2) obtained from SEM is used to calculate the <111> direction etch rate. A sample SEM image of the alignment marks, overall diaphragm and zoomed view of hanging oxide are shown in **Figures 3** and **4a** and **b** respectively. The experiments are performed for six temperatures. The computed <100> and <111> direction etch rates are listed in **Table 1**.

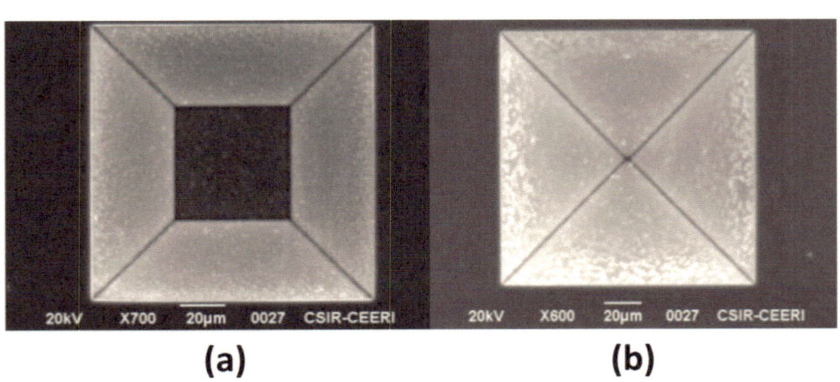

Figure 3.
SEM image of alignment marks used for calculating underetching (a) alignment mark during etching and (b) alignment mark converted to V-groove.

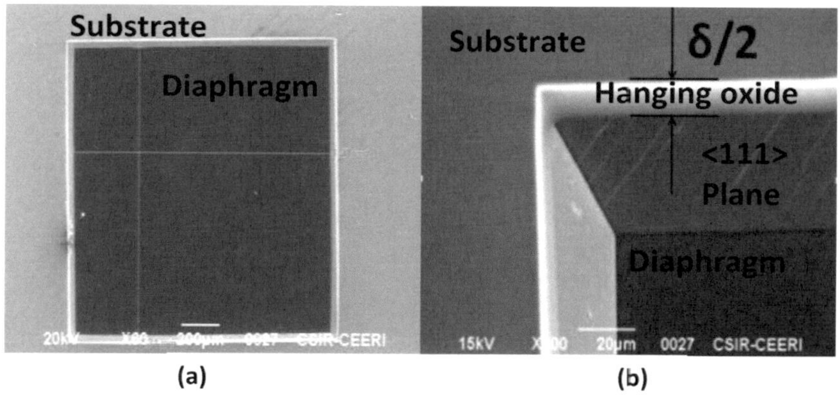

Figure 4.
(a) SEM image of a diaphragm and (b) hanging oxide after TMAH etching [14] copyright 2014 by springer nature (used with permission).

Solution temperature (°C) error = ±1	Etch rate in <100> (µm/h)	Etch rate in <111> (µm/h)	Etch rate ratio <100>/<111>	Underetch per unit depth
63	10.02	1.15	8.71	0.141
68	13.80	1.29	10.69	0.114
73	17.98	1.40	12.84	0.095
78	21.62	1.73	12.49	0.098
83	30.38	1.89	16.07	0.076
88	38.90	2.00	19.45	0.063

Table 1.
Etch rates of Si in TMAH [14]. Copyright 2014 by springer nature (used with permission).

It is evident from the table that the rate of etching increases both in the <100> and <111> direction with increase in temperature. The underetch per unit depth is determined by the ratio of etching in <110> direction to etching in <100> direction. It can be observed that the underetch per unit depth decreases with increased temperature. Alternatively stated, for the same etch depth in <100> direction, lesser underetching is obtained at higher temperatures. A minor deviation in this upward trend is observed between temperature of 73 and 78°C. This may be attributed to some error in experimentally observed data due to temperature variation in the TMAH solution. It may also be attributed to error in the measurement of the diaphragm size from SEM images. The plot of etch rate in <100> direction and underetch per unit depth with solution temperature is shown in **Figures 5** and **6**, respectively.

Based on the above observations, the following empirical equations (linear fit) are proposed for the etching rates in <100> and <111> direction in 25% wt. aq. TMAH at different temperatures:

$$<100>etch\ rate = 1.13 \times T{-}63.21 \quad (4)$$

$$<111>etch\ rate = 0.0365 \times T{-}1.176 \quad (5)$$

A new empirical linear fit equation is obtained from the data in **Table 1** to relate the underetch per unit depth of etching in (100) plane to the solution temperature, which is given by Eq. (6):

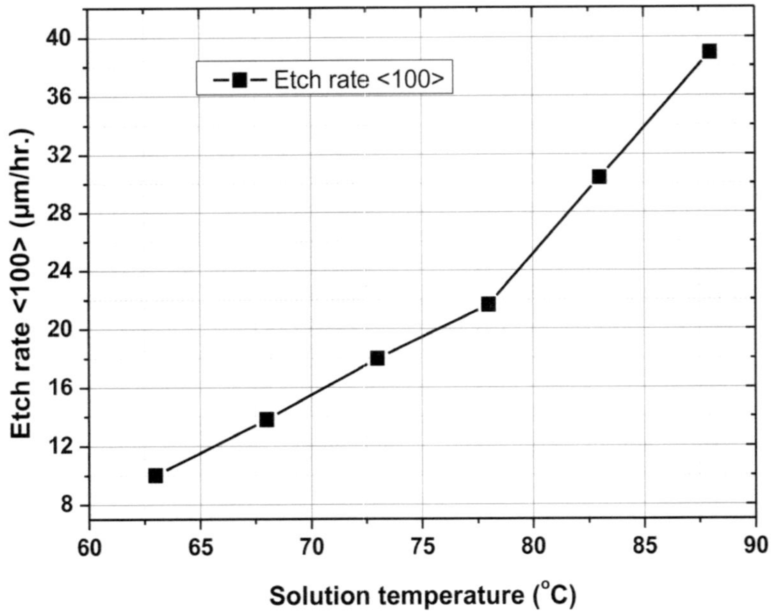

Figure 5.
Plot of etch rate in (100) plane vs. solution temperature.

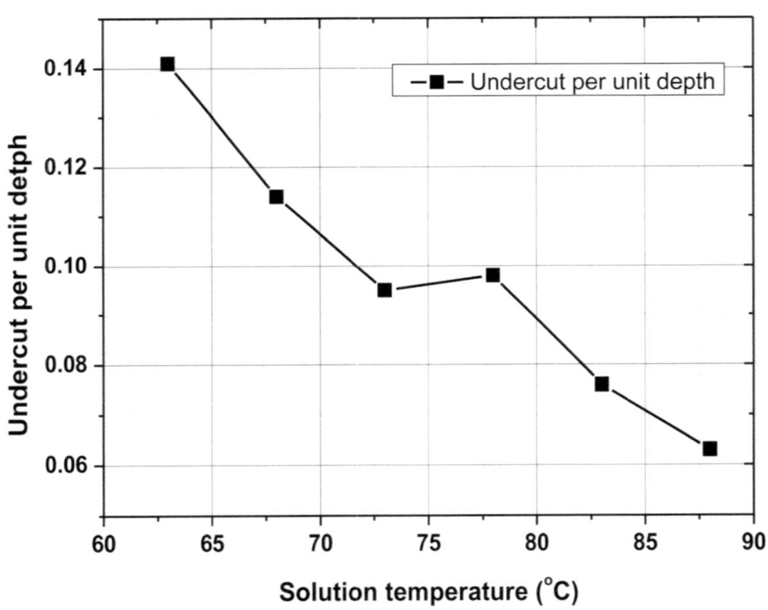

Figure 6.
Plot of underetch per unit depth vs. solution temperature.

$$\text{Underetch per unit depth} = 0.31398\text{–}0.00286 \times \mathrm{T} \tag{6}$$

where T is the solution temperature.

This expression can be used to estimate the dimensions of the masks to be used for obtaining patterns using TMAH etching during fabrication of the pressure sensor. This study provides an important guideline towards making silicon diaphragms with precise dimensional control. Also, the study indicates that in the case of mask sets which have been fabricated without considering the effect of underetching, a higher temperature must be used to minimize the effect of underetching on diaphragm dimensions.

4. Corner compensation structures for TMAH

In many applications of silicon microsensor/actuator structures, like accelerometer and bossed diaphragm pressure sensors, it is desired to have truncated pyramid or mesa type of structures to be realized and/or integrated with thin membrane or beams. During the wet chemical anisotropic etching, it is observed that the convex corners of mesa type structures deteriorates very fast and results in undesired shape/structure. This deterioration is mainly due to the fact that different crystal planes intercept at the convex corner and some of the planes are fast etching planes [10, 16]. The fast etching planes dominate over other crystal planes and hence results in the deterioration of the convex corner shape. This phenomenon is referred to as convex corner undercutting. In sensors structures, like accelerometer, it is needed to preserve the shape of mesas/truncated pyramid. These convex corners can be preserved by adding extra structures (known as convex corner compensation structures) at these convex corners, which are removed or etched out during the process of etching.

Some of the most common corner compensation structures are <110> square, <100> bar type (thin and wide bar structures) and triangle shape structures [17–23]. These compensation structures are shown in **Figure 7**, along with design parameters. Square compensation structures have edges aligned to <110> wafer flat direction is attached to one of the convex corners. Two kinds of bar structures (thin bar and wide bar or bar1 and bar2), aligned in <100> direction, are shown on

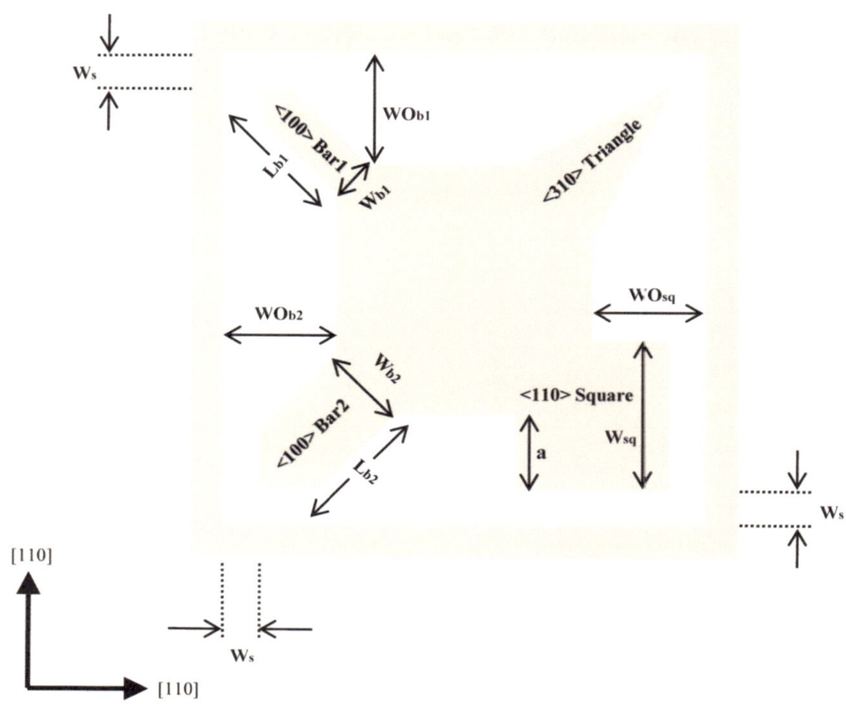

Figure 7.
Corner compensation structures.

two other corners. The convex corner, while the <310> triangle corner compensation structure is shown at the fourth corner. Design parameters and equations of the structures are discussed further in the chapter. In this chapter, we have focused only on <100> bar type structures because these are the ones which can give perfect convex corners with less space. The other structures are reported elsewhere [15].

5. Experimental details

Etch profile or undercutting of the convex corners depends on the crystallographic orientation of the wafer, alignment of structure pattern with the crystallographic direction, type of dopant and its concentration, type of etchant/chemical, chemical concentration, and temperature. We have performed experiments with silicon (100) substrate using 25 wt.% TMAH water solution at 90 ± 1°C temperature in a constant temperature bath. The details of experiment parameters are given in **Table 2**. Thermally grown silicon dioxide has compressive residual stress of ∼100 MPa, which gives better adhesion and reduces the shape deterioration near the mask [24]. After carrying out thermal oxidation ∼1 μm, the mask structures, with different compensating structures as shown in **Figure 7**, were transferred to the silicon substrate with its primary flat aligned with <110> and experiments were performed.

To understand the etch profile and morphology, the samples were removed from the bath, cleaned and examined periodically. Depth was measured using a digital microscope with an accuracy of ±2 μm. It was observed that the profile is consistent in repeated experiments and has a front etch angle of ∼25°, which corresponds to the {311} planes and [310] directions. A square mask aligned to <110> direction without any compensation structure was used to measure the front etch attack angle. The measured data agrees with the reported literature [25, 26]. In KOH {411} planes are responsible for convex corner undercutting [17], while in TMAH {311} planes are responsible [25, 26].

Parameters	Value
Substrate and size	Silicon, 4-in.
Orientation	(100)
Type and dopant	n-type, phosphorus
Resistivity	8–10 Ω-cm
Structure alignment	<110> direction
Substrate thickness	∼525 μm
Masking layer and thickness	Silicon dioxide, ∼1 μm (thermally grown)
Etchant	25 wt.% TMAHW
Temperature	90 ± 1°C
Stirring	No
Etch front angle	24–25°
Fast etch plane	{311}
Fast etch direction	<310>

Table 2.
Silicon substrate specifications and etching parameters [23].

From the experiments the etching parameters were extracted and summarized in **Table 3**, along with compensating structure dimensions.

5.1 <100> thin bar structure etching mechanism

<100> thin bar (structure 1) is longer and has less width. This structure has etching in two sides in <100> direction and from front side in <310> direction. In <100> direction the etching is similar to the (100) plane and in front side it is faster due to fast etch plane {311}. In this compensation structure, the depth is controlled by the width of the bar (twice of the etch depth) and the convex corner shape is preserved by front of the beam. The length should be more than the width. Thus, the convex shape is preserved completely. The convex corner of the final structure obtained by bar1 compensation structure is shown in **Figure 8**.

5.2 <100> wide bar structure etching mechanism

<100> wide bar (structure 2) is shorter in length and wider in width. This structure has also etching in two sides in <100> direction and from front side in <310> direction. In <100> direction the etching is similar to the (100) plane and in front side it is faster due to fast etch plane {311}. In this compensation structure, the depth is controlled by the length of the bar and the convex corner shape is preserved by {310} etch front of the beam. By increasing the width, requirement of the length is reduced. Thus, the convex shape is preserved completely. The convex corner of the final structure obtained by bar2 compensation structure is shown in **Figure 9**.

Parameters	Value
(100) plane etch rate: ER (100)	0.67 μm/min
(111) plane etch rate: ER (111)	0.048 μm/min
(311) plane etch rate: ER (311)	1.494 μm/min
Anisotropic ratio: ER (111)/ER (100)	0.071
Anisotropic ratio: ER (311)/ER (100)	2.23
Minimum side separation: Ws	10 μm (for all the structures)
<100> thin bar structure design dimensions:	
<100> thin bar width: Wb1	860 μm
<100> thin bar length: Lb1	1918 μm
<100> thin bar minimum window opening: WOb1	1366 μm
Alignment angle with <110> primary flat	45°
<100> wide bar structure design dimensions:	
<100> thin bar width: Wb2	1802 μm
<100> thin bar length: Lb2	1499 μm
<100> thin bar minimum window opening: WOb2	1070 μm
Alignment angle with <110> primary flat	45°

Table 3.
Etch rate parameters and compensating structure dimensions [23].

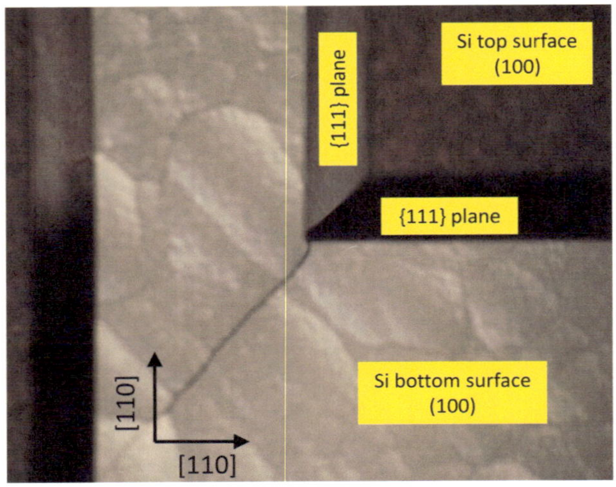

Figure 8.
Optical photograph of the convex corner realized using the <100> thin bar. [23] Copyright 2006 by IOP Publishing Ltd. (used with permission).

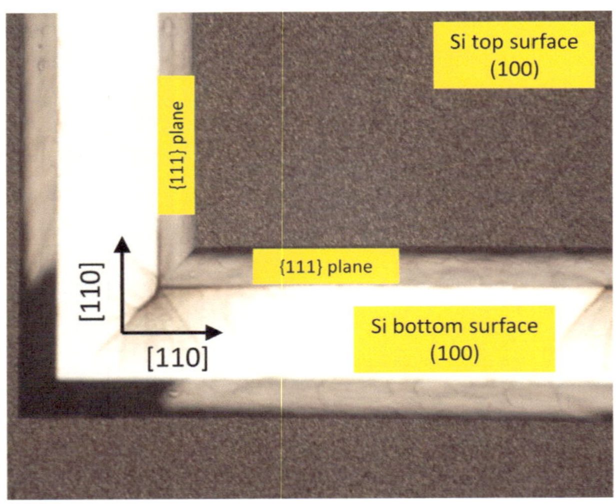

Figure 9.
Optical photograph of the convex corner realized using the <100> wide bar. [23] Copyright 2006 by IOP Publishing Ltd. (used with permission).

6. Design analysis and discussions

Optical photographs of the realized convex corners using the <100> thin and wide bars are depicted in **Figures 8** and **9**. As shown in **Figure 8**, for the dimensions mentioned earlier, a perfect convex corner is obtained for the <100> thin bar with free end at a depth of etching equal to 430 μm. As shown in **Figure 9**, for the dimensions mentioned earlier, a perfect convex corner is obtained for the <100> wide bar at a depth of etching equal to 485 μm. The thin bar is narrower in terms of width and longer in terms of length than the wide bar structure.

We can infer from **Figures 8** and **9** that perfect convex corners are obtained in each of the compensation structures—wide bar and thin bar. Bar2 requires less window opening compared to bar1. On the basis of the examination of the experimental results, equations to design these structures are deduced and reported in Mukhiya et al. [23]. For bar1 structure, design is given by Eqs. (7) and (8), as follows [23]:

$$W_{b1} = 2D_e \tag{7}$$

$$WO_{b1} = W_s + 1.414\frac{R(311)}{R(100)}D_e \tag{8}$$

For bar2 structure, design is given by Eqs. (9) and (10), as follows [23]:

$$W_{b2} \le \left[2.07\,\frac{R(311)}{R(100)} - 0.336\right]D_e \tag{9}$$

$$WO_{b2} \ge W_s + \left[0.351 + 0.734\,\frac{R(311)}{R(100)}\right]D_e \tag{10}$$

where D_e is the etch depth. The remaining symbols have been defined earlier in the chapter. In order to obtain perfect convex corners for the bar structures, Eqs. (7)–(10) provide the biggest and smallest dimensions of width of beam and window opening, respectively.

From etch profile of both the <100> bar compensation structures, it is observed that they self-align with the [310] direction, which is the fastest etch plane direction. Compensating structure designed with [310] triangle shape will have consistent etch profile but it will require more space. Bar2 structure is ~37% more space efficient as compared to bar1 for the same etch depth.

7. Conclusions

With the advancement of technology, it is preferred to have CMOS-MEMS integration. This imposes the challenge of development of CMOS compatible processes and micromachining techniques. Among various micromachining techniques, wet bulk micromachining is a preferred technique because it is easy, cost-effective and has well defined characteristics. In spite of extensive research in this field, wet bulk micromachining using TMAH is still an interesting area of research among the researchers and academicians. Post process CMOS compatible wet bulk micromachining is possible with TMAH. By adding some additives, it is also possible to protect Aluminum in TMAH-based etchants. In the presented work, etch rates for different crystal planes have been measured at various temperatures. It is observed that for 25 wt.% TMAH underetching of {111} planes increase with increase in temperature. However, the anisotropy ratio (111/100) decreases with temperature. Etch rate of both the crystal planes, (100) and (111), increases with temperature, as bond becomes weak at higher temperature. Etch rate of (100) plane is faster than the etch rate of (111) because (100) is a low atomic density plane. Empirical design equations have been derived for the etch rates as a function of temperature. As a use case of this anisotropic etching, diaphragm of pressure sensor has been fabricated at the authors' laboratory and its analysis in context of TMAH etching is presented.

It is observed that during wet anisotropic etching using 25 wt.% TMAH, the convex corners etches very fast and deteriorates in shape. This deterioration is known as undercutting, and it is mainly due to the fact that different crystal planes are encountered at the convex corner and some of the crystal planes etch very fast in comparison to other planes. These fast etching planes are responsible for the undercutting. For TMAH, the fast etching planes are found at an angle of ~25°, which are {311} plane. In case of TMAH, the underetching and undercutting are found faster than its counterpart KOH. In many applications, where these convex corners are required to be protected, some additional structures are added to protect the convex corner from deterioration. These are known as convex corner compensation structures. Most common compensation structure shave been discussed and optimum structure, i.e., <100> bar type of structures (thin bar and wide bar) have been discussed in detail. Both the structures can protect the mesa and can give perfect convex corner. Bar2 structure is more space efficient than bar1 structure. Using this type of compensation structures, accelerometer proof mask has been fabricated at authors' laboratory and presented. Generalized empirical design equations have also been discussed.

Acknowledgements

Authors would like to acknowledge the generous support of the Director, CSIR-CEERI, Pilani. The authors would also like to thank all the scientific and technical members of Smart Sensors Area at CSIR-CEERI, Pilani. The financial support by CSIR, New Delhi to carry out the research work through various projects is gratefully acknowledged. R Mukhiya acknowledges FBK, Trento, Italy, for the experimental study on corner compensation structures, and DST for the financial support through the ITPAR project.

Author details

S. Santosh Kumar* and Ravindra Mukhiya
CSIR-Central Electronics Engineering Research Institute, Pilani, Rajasthan, India

*Address all correspondence to: santoshkumar.ceeri@gmail.com

References

[1] Bhat KN. Silicon micromachined pressure sensors. Journal of the Indian Institute of Science. 2007;**87**(1): 115-131

[2] Sharma A, Mukhiya R, Santosh Kumar S, Gopal R, Pant BD. Dynamic characterization of bulk micromachined accelerometer using laser doppler vibrometer (LDV). Microsystem Technologies. 2015;**21**:2221-2232. DOI: 10.1007/s00542-014-2316-3

[3] Wei J, Sarro PM, Duc TC. A piezoresistive sensor for pressure monitoring at inkjet nozzle. In: Proceedings of IEEE Sensors 2010; 1-4 November 2010; USA, New York: IEEE; 2011. pp. 2093-2096

[4] You JS, Kim DH, Huh JY, Park HJ, Pak JJ, Kang CS. Experiments on anisotropic etching of Si in TMAH. Solar Energy Materials and Solar Cells. 2001; **66**:37-44. DOI: 10.1016/S0927-0248 (00)00156-2

[5] Bashir R, Hilt JZ, Elibol O, Gupta A, Peppas NA. Micromechanical cantilever as an ultrasensitive pH microsensor. Applied Physics Letters. 2002;**81**(16): 3091-3093. DOI: 10.1063/1.1514825

[6] Tokoro K, Uchikawa D, Shikida M, Sato K. Anisotropic etching properties of silicon in KOH and TMAH solutions. In: 25-28 Nov. 1998; Japan, New York: IEEE; 2002. pp. 65-70

[7] Yan G-Z, Chan PCH, Hsing I-M, Sharma RK, Sin JKO. An improved Si-etching solution without attacking exposed aluminum. Sensors and Actuators A. 2001;**89**:135-141. DOI: 10.1109/MEMSYS.2000.838579

[8] Fujitsuka N, Hamaguchi K, Funabashi H, Kawasaki E, Fukada T. Aluminum protected silicon anisotropic etching technique using TMAH with an oxidizing agent and dissolved Si. R&D Review of Toyota CRDL. 2004;**39**:34-40

[9] Seidel H. The mechanism of anisotropic silicon etching and its relevance for micromachining. In: Proc. Transducers '87, Rec. 4th Int. Conf. Solid-State Sensors and Actuators; 2-5 June 1987; Tokyo: Japan; pp. 2093-2096

[10] Koide, Sato K, Tanaka S. Simulation of two-dimensional etch profile of silicon during orientation-dependent anisotropic etching. In: Proceedings of the IEEE Microelectromechanical Systems (MEMS) Workshop; Nara: Japan; 1991. pp. 216-220

[11] Tabata O, Asahi R, Funabashi H, Shimaoka K, Sugiyama S. Anisotropic etching of silicon in TMAH solutions. Sensors and Actuators A. 1992;**34**:51-57. DOI: 10.1016/0924-4247(92)80139-T

[12] Pal P, Sato K, Gosalvez MA, Shikida M. Study of rounded concave and sharp edge convex corners undercutting in CMOS compatible anisotropic etchants. Journal of Micromechanics and Microengineering. 2007;**17**:2299-2307. DOI: 10.1088/ 0960-1317/17/11/017

[13] Thong JTL, Choi WK, Chong CW. TMAH etching of silicon and the interaction of etching parameters. Sensors and Actuators A. 1997;**63**: 243-249. DOI: 10.1016/S0924-4247(97) 80511-0

[14] Santosh Kumar S, Pant BD. Design principles and considerations for the 'ideal' silicon piezoresistive pressure sensor: A focused review. Microsystem Technologies. 2014;**20**:1213-1247. DOI: 10.1007/s00542-014-2215-7

[15] Mukhiya R, Bagolini A, Bhattacharyya TK, Lorenzelli L, Zen M. Experimental study and analysis of

corner compensation structures for CMOS compatible bulk micromachining using 25 wt.% TMAH. Microelectronics Journal. 2011;**42**:127-134. DOI: 10.1016/j.mejo.2010.08.018

[16] Offereins HL, Kühl K, Sandmaier H. Methods for fabrication of convex corners in anisotropic etching of (100) silicon in aqueous KOH. Sensors and Actuators A. 1990;**25**:9-13. DOI: 10.1016/0924-4247(90)87002-Z

[17] Mayer GK, Offereings HL, Sandmaier H, Kühl K. Fabrication of non-underetched convex corner in anisotropic etching of (100) silicon in aqueous KOH with respect to novel micromechanical elements. Journal of the Electrochemical Society. 1990;**137**:3947-3951. DOI: 10.1149/1.2086334

[18] Enoksson P. New structure for corner compensation in anisotropic KOH etching. Journal of Micromechanics and Microengineering. 1997;**7**:141-144. DOI: 10.1088/0960-1317/7/3/016

[19] Zhang Q, Liu L, Li Z. A new approach to convex corner compensation for anisotropic etching of (100) Si in KOH. Sensors and Actuators A. 1996;**56**:251-254. DOI: 10.1016/S0924-4247(96)01312-X

[20] Puers B, Sansen W. Compensation structures for convex corner micromachining in silicon. Sensors and Actuators A. 1990;**23**:1036-1041. DOI: 10.1016/0924-4247(90)87085-W

[21] Bao M, Burrer C, Esteve J, Bausells J, Marco S. Etching front control of <110> strips for corner compensation. Sensors and Actuators A. 1993;**37-38**:727-732. DOI: 10.1016/0924-4247(93)80123-X

[22] Wu X-P, Ko WH. Compensating corner undercutting in anisotropic etching of (100) silicon. Sensors and Actuators A. 1989;**18**:207-215. DOI: 10.1016/0250-6874(89)87019-2

[23] Mukhiya R, Bagolini A, Margesin B, Zen M, Kal S. <100> bar corner compensation for CMOS compatible anisotropic TMAH etching. Journal of Micromechanics and Microengineering. 2006;**16**:2458-2462. DOI: 10.1088/0960-1317/16/11/029

[24] Takao H, Yong C-C, Rajanna K, Ishida M. Shape deterioration of mesa structures in post-CMOS anisotropic etching of silicon microstructures: An experimental study. Sensors and Actuators A. 2000;**86**:115-121. DOI: 10.1016/S0924-4247(00)00437-4

[25] Tellier CR, Charbonnieras AR. Characterization of the anisotropic chemical attack of (h h l) silicon plates in a TMAH 25 wt.% solution: Micromachining and adequacy of the dissolution slowness surface. Sensors and Actuators A. 2003;**105**:62-75. DOI: 10.1016/S0924-4247(03)00064-5

[26] Trieu HK, Mokwa W. A generalized model describing corner undercutting by the experimental analysis of TMAH/IPA. Journal of Micromechanics and Microengineering. 1998;**8**:80-83. DOI: 10.1088/0960-1317/8/2/009

Chapter 5

Physical Processes and Plasma Parameters in a Radio-Frequency Hybrid Plasma System for Thin-Film Production with Ion Assistance

Elena Kralkina, Andrey Alexandrov, Polina Nekludova, Aleksandr Nikonov, Vladimir Pavlov, Konstantin Vavilin, Vadim Odinokov and Vadim Sologub

Abstract

The results of the study of the plasma reactor on the combined magnetron discharge and radio-frequency (RF) inductive discharge located in the external magnetic field are presented. Magnetron discharge provides the generation of atoms and ions of the target materials, while the flow of accelerated ions used for the ion assistance is provided by the RF inductive discharge located in an external magnetic field. Approaching the region of resonant absorption of RF power by optimizing the magnitude and configuration of the external magnetic field makes it possible to obtain a uniform within 10% radial distribution of the ion current across the diameter of 150 mm. When the RF power supply power is 1000 W, the ion current density on the substrate can be adjusted in the range of 0.1–3 mA/cm^2. The use of ion assisting results in a fundamental change in the structure and properties of functional coatings, deposited using a magnetron.

Keywords: RF inductive discharge, assisting ions, helicon, Trivelpiece-Gould wave, magnetic field, magnetron, film deposition

1. Introduction

At the present time, vacuum plasma methods for the formation of multicomponent thin-film structures based on magnetron or vacuum arc gas discharges are widely used in the industry. These methods allow to obtain a wide class of functional coatings, such as optical, hardening, anticorrosion, antibacterial, etc. A particular case of the known vacuum plasma methods of coating formation is ion-assisted deposition [1–8]. The method involves continuous or periodic bombardment of growing thin films by accelerated ions.

The results of the bombardment of the substrate and the growing films by accelerated ions are [1–8]:

- cleaning of the substrate surface and removing traces of water and hydrocarbons;

- increase of density and, in some cases, modification of the growing film structure;

- removal of loosely bounded molecules during the film growth;

- improving the adhesion of the film to the substrate;

- better control of the film stoichiometric composition in the case of chemically active gas usage.

The concept of ion-stimulated deposition was proposed more than 70 years ago by Berghaus [1]. More than two decades later, the technology was implemented in the experimental works by Mattox and MacDonald [2], as well as by Mattox and Kominiak [3].

Currently, vacuum deposition of films is carried out by way of thermal evaporation, magnetron, vacuum arc, and ion-beam deposition [4–8]. Bombardment of substrates and growing films is implemented using accelerated ion beams, generated in ion sources or gas-discharge plasma. In the latter case, a negative bias is applied to the substrate, which accelerates the ions in the direction perpendicular to the surface of the growing film.

The effect of the ion flux treatment on the thin-film properties substantially depends on the flux magnitude, energy and mass of the ions, as well as the ratio between the flows of the assisting ions and the atoms of the deposited substance. The question of the optimal magnitude of energy, which has to be introduced into the growing film per a single deposited atom, was examined in [5, 9–12].

In [10], the results of numerous experiments on the deposition of coatings using ion beam stimulation are analyzed. It is shown that the most significant changes in the properties of deposited films occur, when each of the deposited atom obtains additional energy in the range of 1.0–100 eV. However, it was shown in [11] that in the case of Ti-Al-N film deposition, composition of the film, predominant orientation, and distances between atoms in the lattice significantly depend on whether the energy is introduced by one high-energy ion or by several low-energy ions. As shown in [12], the problem is that in the above cases the energy is transferred to the deposited atoms irregularly. High-energy ions are capable of creating defects in the deeper layers of the coating and the substrate; having been reflected from the surface of the substrate, the ions acquire energy, which significantly depends on the substrate material. This results in the fact [12] that the energy required for the film growth at the initial stage may greatly differ from the energy, which is required at the final growth stage, when the assisting ions interact only with the atoms of the deposited substance. It should be noted that when the ion energy exceeds 15–30 eV [13], the growing film can be sputtered with a beam of fast ions. The use of ion beams with energies up to 30 eV makes it possible to reduce these effects to a minimum.

The use of vacuum plasma methods of forming functional coatings in the industry requires the ensuring of high-speed film deposition and, accordingly, the use of significant fluxes of assisting ions. Literature review [6, 13] shows that the ion fluxes generated by gridded and gridless ion beam sources are often insufficient to assist the growth of films produced using magnetron and vacuum arc methods. Besides, there is a complex problem of matching the operation of gas-discharge sources generating assisting ions and coating systems in terms of pressure.

In [14], the possibility of generating plasma flows with the ion component density of up to 20–30 mA/cm^2 and independently controlled ion energy within the range of 20–120 eV by using a combination of arc discharge with inductive radio-frequency (RF) discharge in external magnetic field is shown. The results obtained in [14] served as the starting point for the development of a plasma reactor [15–18], intended for the magnetron sputtering of functional coatings using stimulation by ions, generated in RF inductive discharge with external magnetic field.

2. Basic physical patterns underlying the design of a plasma reactor

The task of developing a plasma reactor for magnetron sputtering deposition of ion-stimulated coatings has determined the composition of the reactor: one or more magnetron sources to provide the flow of sputtered particles onto a substrate and gas-discharge source, which provides the generation of a magnitude-controlled flux of ions, bombarding the substrate. The ion flux energy control is provided by a direct current (DC) or RF biasing of the substrate.

To enable successful operation of the plasma reactor, it is necessary to ensure that:

1. The working areas of the magnetron and gas-discharge sources are matched in terms of pressure.

2. Changing of external parameters of the gas-discharge source ensures coordination of the flux of assisting ions and the flux of sputtered particles entering the substrate.

3. Changing of external parameters of the gas-discharge source ensures uniform density of ion current density on the substrate not worse than 10%.

4. The plasma reactor operates using both inert and chemically active gases.

Let us do some estimates. Taking into account typical rate of magnetron sputtering, it is advisable to ensure that the density of the stimulating ion current varies within the range of 0.1–3 mA/cm^2. This means that the plasma density must vary within the range of $0.03–1 \times 10^{11}$ cm^{-3}.

A literature review [19–21] has shown that it is reasonable to consider two modifications of inductive RF discharge as the options for implementing the working process in the gas-discharge source, namely, inductive RF discharge in the absence and in the presence of external magnetic field. In the first case, the magnitude of the assisting ion flow will be determined by the power of the RF power supply. In the second case, two external parameters—the power of the RF power supply and the induction of an external magnetic field—will allow to control the values of plasma parameters and their spatial distribution in the plasma reactor.

In modern industrial installations, the diameter of substrates usually exceeds 200 mm. Large diameter of the plasma reactor makes it difficult to generate strong magnetic fields in the volume of the reactor. It is known [22–25] that when an external magnetic field with induction of less than 100 Gs is imposed on inductive RF discharge, a peak of electron density is observed at certain resonant values of the magnetic field. The absolute values of plasma density are close to those required in case of our task. The physically observable density peak is associated with resonant excitation in the plasma reactor of helicon and Trivelpiece-Gould waves [22–25]. Provided that the pressure in the reactor does not exceed 10 mTorr, the

Trivelpiece-Gould wave is a bulk wave, penetrates deep into plasma, and determines the absorption of RF power [24, 25].

In [26], the authors offered a scheme of two-chamber process plasma source, based on inductive RF discharge with external magnetic field, diverging toward the substrate. Following the scheme [26], preliminary experiments on the study of the discharge parameters have been carried out with the plasma reactor having two chambers [27]. The model consisted of two cylindrical quartz chambers of different diameters and a magnetic system. The upper part of the reactor (gas-discharge chamber) had a diameter of 10 cm and a height of 25 cm. The lower part of the reactor (process chamber) had 46 cm diameter and 30 cm height. The RF power input unit—a three-turn solenoid antenna—was located on the outer surface of the gas-discharge chamber. The magnetic system allowed generation in the volume of the process chamber of uniform magnetic field, converging and diverging toward the lower flange where the substrate was located.

Experiments have shown that when using an inductive RF discharge in the absence of magnetic field, the discharge is concentrated in the gas-discharge chamber, while the plasma density in the process chamber near the substrate is vanishingly small. A different situation is observed when using external magnetic field, diverging in the region of the gas-discharge chamber and uniform in the region of the process chamber. Under argon pressures, when the electron mean free path exceeds the longitudinal size of the system, the imposition of magnetic field results in significant changes in the discharge length [27]. In the absence of magnetic field, the discharge is concentrated in the gas-discharge chamber. An increase in the magnitude of the magnetic field first results in the appearance of plasma in the upper part of the technological chamber. Then, the length of the intensely glowing part of the discharge in the process chamber starts to grow, and, finally, the discharge closes at the bottom flange. The movement of electrons across the magnetic field is difficult; therefore, an extended plasma column, sharply outlined in the radial direction, appears in the process chamber. In parallel

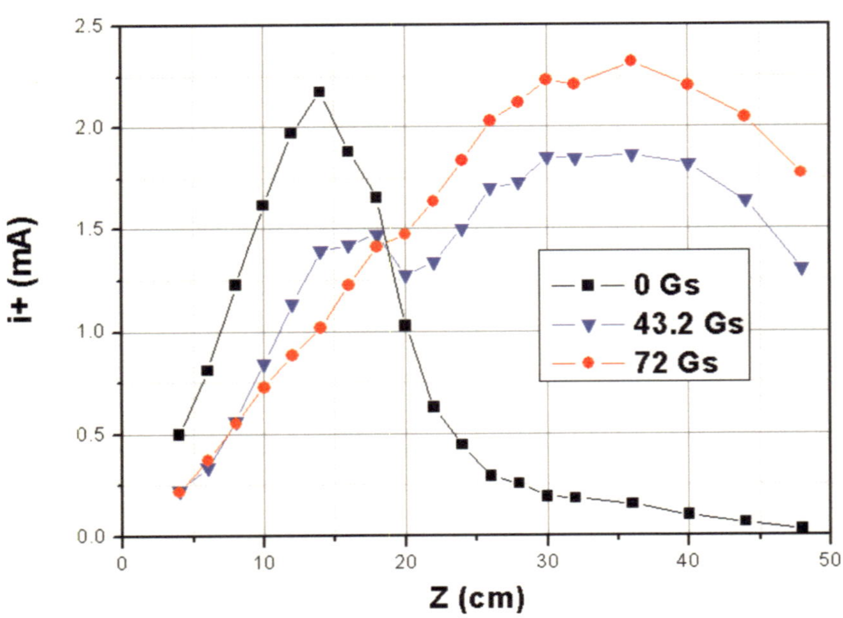

Figure 1.
Probe ionic saturation current distribution along z axis of plasma source depending on external magnetic field magnitude: 0.7 mTorr, 400 W, and 13.56 MHz.

with the increase in the discharge length near the substrate, an increase in the probe ion saturation current is observed.

Figure 1 shows variation of the axial distribution of the probe ion saturation current when the magnetic field induction grows. In the absence of magnetic field, the discharge is concentrated in the upper part of the plasma source; as the magnetic field increases, the ion current in the lower chamber gets increased. When the magnetic field exceeds 36 Gs, the discharge gets localized in the process chamber of the reactor.

This effect is observed for all of the examined powers of the RF power supply, and the maximum values of the ion current are increasing in proportion to the input power. The achieved plasma concentrations near the substrate correspond to those required in this work. Detailed studies of the physical causes for the axial redistribution of plasma density as the magnetic field induction increases have shown that the effect is associated with the patterns of excitation of partial standing waves in the plasma [17, 27]. It should be noted that the use of other configurations of the magnetic field did not result in significant increase in the ionic saturation current near the substrate.

The obtained results served as the basis for the development of a semi-industrial installation for magnetron deposition of coatings with ion assistance.

3. Design of a plasma reactor, which combines magnetron and inductive RF discharge with external magnetic field

Preliminary results, obtained using a plasma reactor prototype, have served as the basis for a semi-industrial installation [16–18]. Diagram of this prototype is shown in **Figure 2**. The reactor consists of two parts. The main part is a metal cylindrical process chamber having 500 mm diameter and 350 mm height. At the bottom of the chamber, there is a rotating table for the accommodation of samples being treated. To facilitate spectrometric studies of plasma parameters, two optical inspection windows are located above the table right opposite to each other. The magnetron source is installed on the side surface of the process chamber.

Quartz gas-discharge chamber having 250 mm length and 220 mm diameter is mounted on top of the process chamber. From above the chamber, volume is covered with a glass blind flange; from beneath the chamber is sealed with a metal flange with an opening that allows plasma to penetrate into the main chamber.

Below the process chamber, a pumping system is installed, consisting of rotary and turbo molecular pumps. The working gas is supplied to the reactor through the gas inlet located in the upper part of the gas-discharge chamber.

The magnetic system consists of two electromagnets, located in the upper and lower parts of the process chamber. The electromagnets ensure generation of diverging magnetic field in the area of the gas-discharge chamber. The magnitude of the magnetic field in each specific point in the volume of the process chamber is determined by the currents I_{top}, I_{bot} flowing through the top and bottom electromagnets, respectively, and one and the same value of magnetic induction can be provided by setting different ratios between the currents of the magnets. In this regard, in the further section, the dependence of the discharge parameters on I_{top}, I_{bot} is given in the graphs illustrating the results of experiments. **Table 1** contains the magnetic field induction values achieved near the antenna, in the center of the process chamber and near the substrate at different values of I_{top}, I_{bot}.

To excite an inductive RF discharge, a solenoid antenna is used, which is located on the outer surface of the quartz chamber. The ends of the antenna are connected

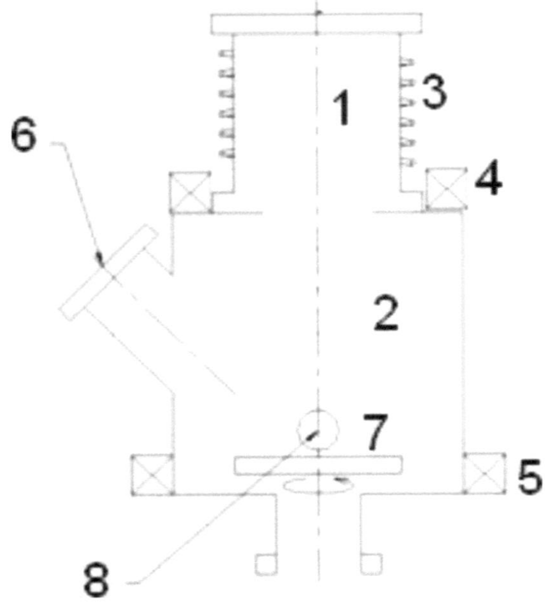

Figure 2.
Plasma reactor prototype diagram: (1, 2) gas-discharge and process chambers, (3) antenna, (4, 5) top and bottom electromagnets, (6) magnetron, (7) rotating table, and (8) window.

Top magnet current (Gs)	Bottom magnet current (Gs)	Magnetic field in the substrate plane (Gs)	Magnetic field in the bottom part of the gas-discharge chamber (Gs)
3	5	39	49
4	5	40	63
5	5	42	78
7	5	45	107
3	1	11	45
3	3	25	47
3	7	53	50
3	9	66	52
3	12	87	55
10	4	42	150
1	6	42	20

Table 1.
Correlation between the currents of the magnets and the magnitude of the magnetic field near the substrate and in the gas-discharge chamber.

through the matching system to the RF power source, having the operating frequency of 13.56 MHz and an output power of up to 1000 W. In the inductive RF discharge, the power of RF power sources is coupled not only to plasma but is wasted in the antenna too. In order to measure the RF power value P_{pl} coupled to plasma, the method described in [24] was used.

To ignite and maintain the magnetron discharge, the RF power source with the operating frequency of 13.56 MHz and output power up to 1000 W is used.

To study the homogeneity of plasma near the substrate, a movable Langmuir probe was used, capable of traveling along the diameter of the substrate. When measuring the dependence of the probe ionic saturation current on the magnitude and configuration of the magnetic field, 60 V potential negative in respect of to the walls of the main chamber is applied to the probe.

In parallel with probe measurements, the plasma glow spectrum was studied. Plasma radiation via an optical fiber was fed to the input of the MDR-40 monochromator, on the output of which photomultiplier FEU-100 was installed. The signal from the photomultiplier was amplified and fed to an ADC connected to a computer. The spectrum was scanned in the wavelength range of 400–700 nm. In addition to the glow spectrum, longitudinal distribution of the spectral lines intensity was measured. To this end, the light guide was moved along the generating line of the process chamber within the diagnostic window opening.

Thin films were deposited onto silicon substrates when only the magnetron was operating and when the magnetron and the gas-discharge plasma source were operating together in the absence and in the presence of bias on the substrate. The morphology of the surface and the cleavage of the films were studied with the help of scanning electron microscope Supra-40. The electro-physical characteristics of the coatings were measured with the help of a two-probe method.

4. Results of the gas-discharge plasma source studies.

At the first stage of the study, the parameters of the plasma generated by inductive RF discharge were studied depending on the size and configuration of the external magnetic field, the power of the RF power supply, and the pressure of argon in the absence of the magnetron discharge.

Experimental studies of the discharge have shown that when a longitudinal magnetic field is imposed on the discharge, an extended plasma column is formed in the process chamber, similar to that observed in the prototype reactor. The diameter of the plasma column is approximately equal to the diameter of the quartz chamber that is 20 cm (**Figure 3**).

As with the prototype, a change in the magnetic field made it possible to control the longitudinal distribution of plasma density. **Figure 4** shows correlation between the plasma glow intensity I_1, measured in the central part of the process chamber, and at the substrate I_3. An increase in the currents through the magnet, located near the gas discharge I_{top} and the substrate I_{bot}, results in equalization of the distribution of the plasma glow intensity along the discharge axis. This confirms the conclusion about the formation of the plasma "column," closing on the substrate.

Figures 5 and **6** show the dependence of the portion of the power P_{pl} coupled to the argon plasma, on the values of the currents I_{top}, I_{bot} through the top and bottom magnets, provided that the power of the RF power supply is 1000 W. As can be seen, P_{pl} is determined by the magnetic field B, which is generated by the current through the top electromagnet located in the area of the gas-discharge chamber. The effect induced by the bottom magnet is noticeable at low currents of the top magnet, when contribution of the bottom electromagnet to the values of the magnetic field induction is significant. The non-monotonic character of dependence of the absorbed power on B is also worth noticing. P_{pl} reaches the maximum in the I_{top} range between 3 and 5 A, which corresponds to magnetic fields in the area of the gas-discharge chamber of 50–80 Gs. The non-monotonous nature of the dependence of the absorbed power on the external magnetic field is retained when argon is replaced by other working gases, such as neon, argon, and oxygen. However, the position of the energy input maximum significantly depends on the working

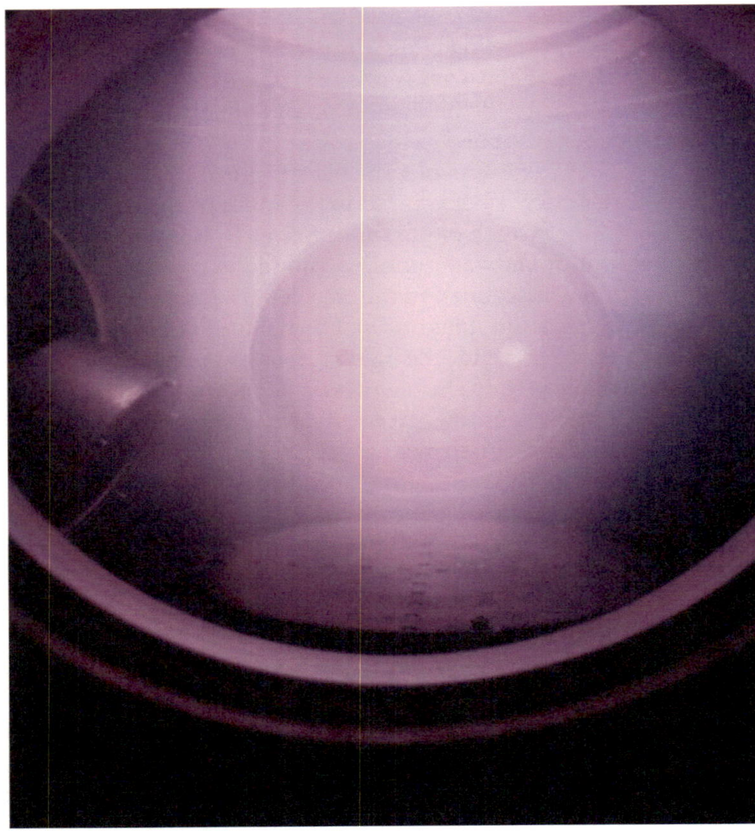

Figure 3.
Photo of the discharge in the plasma reactor.

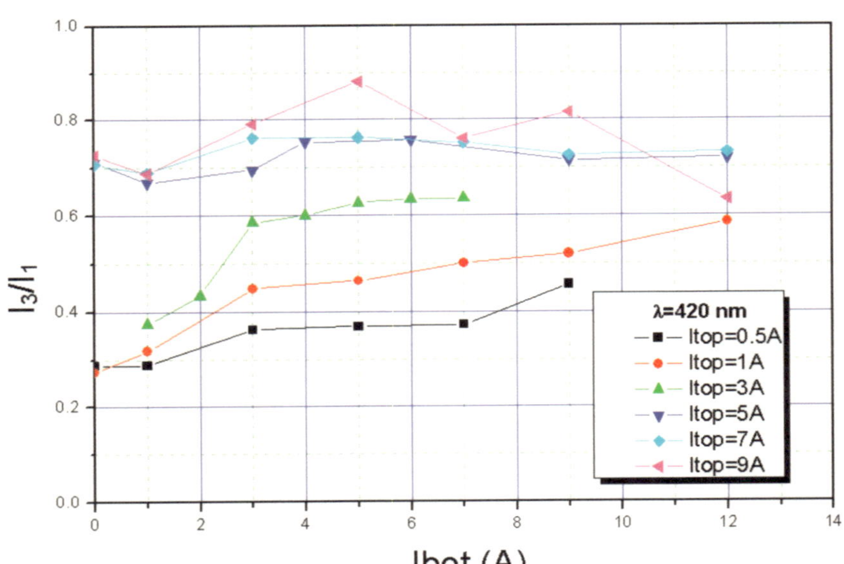

Figure 4.
The ratio of plasma glow intensity near the substrate I_3 to the plasma glow intensity near the magnetron I_1 versus the current flowing through the bottom electromagnet. Corresponding values of the current through the top electromagnet are shown in the menu in the figure. Pressure, 0.2 Pa.

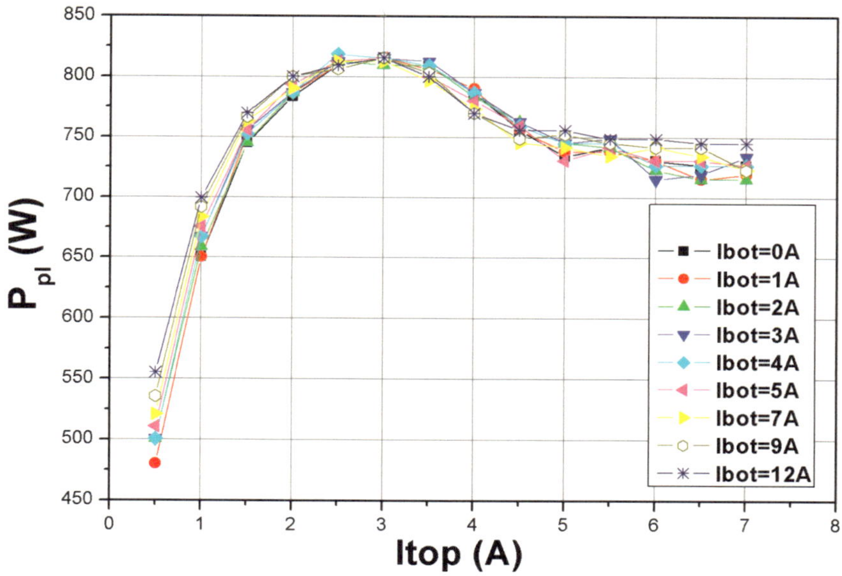

Figure 5.
Dependence of the portion of power, coupled to plasma, on the values of currents through the top and bottom magnets, provided that the RF power supply power is 1000 W. Argon pressure, 0.7 Pa.

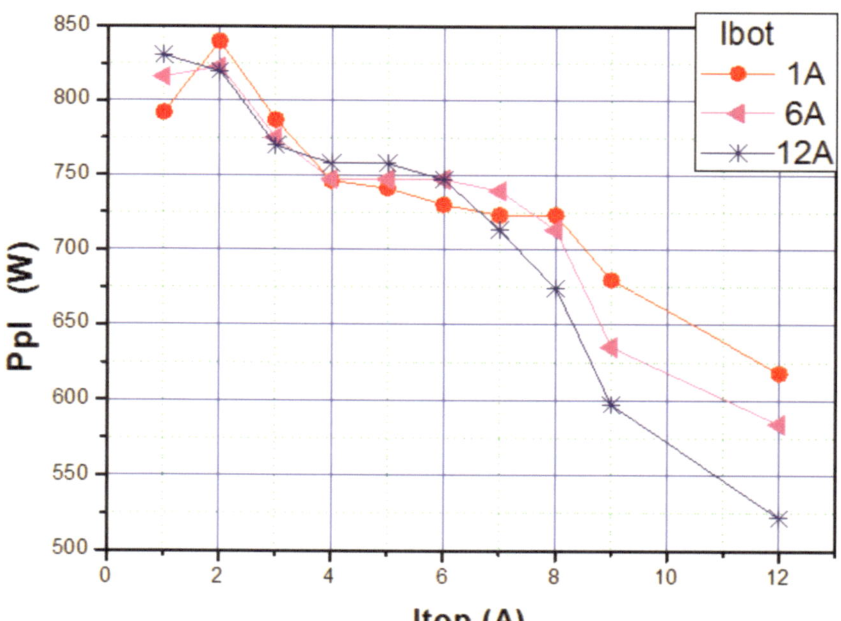

Figure 6.
Dependence of the portion of power, coupled to plasma, on the values of currents through the top and bottom magnets, provided that the RF power supply power is 1000 W. Oxygen pressure, 0.7 Pa.

gas used. Thus, with the use of oxygen, position of the $P_{pl}(B)$ maximum at the RF power supply power of 1000 W is reached at I_{top} = 2 A (see **Figure 6**).

Figure 7 shows radial dependence of the RF longitudinal magnetic field B_z, measured at various I_{top} values. One can see that B_z under condition of the best absorption of the RF power (at the magnetic field of about 80 G in the area of the

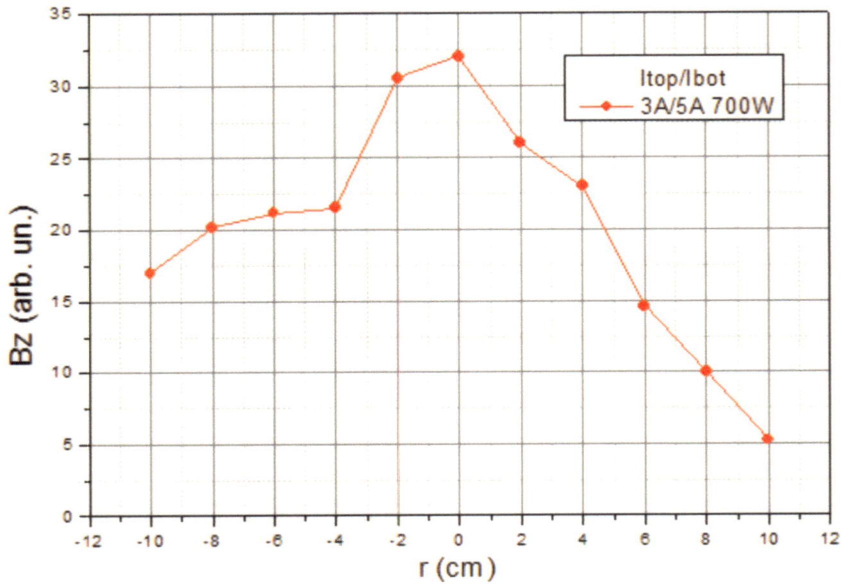

Figure 7.
Radial dependence of the RF longitudinal magnetic field.

gas-discharge chamber) reaches its maximum on the discharge axis, that is, the RF fields penetrate deep into the plasma (see **Figure 7**). This indicates excitation of bulk waves in the discharge.

Calculation of the dependence of the RF fields, excited in plasma, on the external magnetic field, made on the basis of the theoretical model of a limited inductive plasma source with external magnetic field [25], has shown that interconnected helicons and Trivelpiece-Gould waves are excited in the discharge under considered experimental conditions. The Trivelpiece-Gould wave is a bulk wave, and it is the dissipation of its energy that determines the absorption of RF power in the plasma.

The conclusion concerning the resonant excitation of bulk waves at the current of the top magnet of 5A is also confirmed by measurements of the probe ionic saturation current near the substrate (see **Figure 8**). An increase in the current of the top magnet within the range from 1 to 5 A results in increase in the absolute values of the ionic current and the formation of a bell-shaped distribution. As the current is further increased to 7 A, the portion of RF power absorbed by the plasma and the absolute value of the probe ionic saturation current decrease. It is necessary to note the result that is important for technological applications: at the top magnet current of 3A, the most uniform distribution of the ion current is observed. Deviation of the ionic current values from the average ones within the 20 cm diameter does not exceed 10%. Remarkably, the most uniform radial distribution near the substrate is observed not in the resonance region, but when approaching the resonance region.

The most uniform distribution, obtained at the top magnet current of 3A, corresponds to the ionic current density of about 1 mA/cm^2. This value may be insufficient for ionic assistance at high rates of film deposition. Additional experiments have shown that increasing the bottom magnet current allows to increase the ionic current density by two to three times and to obtain homogeneous plasma within the diameter of 15 cm (**Figure 9**, curves 1 and 2).

Figure 9 additionally shows the radial distributions of the ionic saturation current, obtained at the top magnet current of 1A. As a reminder, the point I_{top} = 1 A is located before the maximum of the $P_{pl}(I_{top})$ curve. It is natural to assume that at this

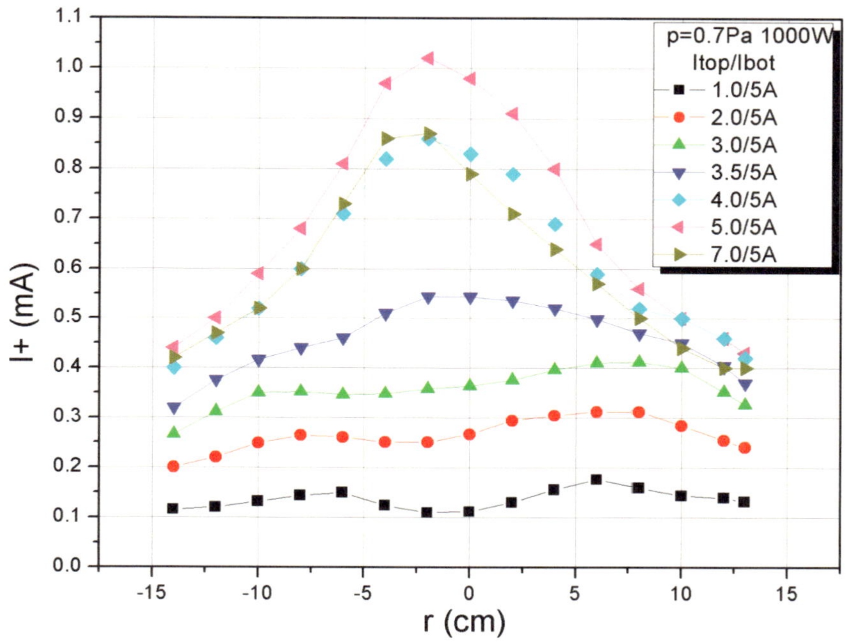

Figure 8.
Radial dependence of ionic saturation current at different top magnet currents while the bottom magnet current is fixed.

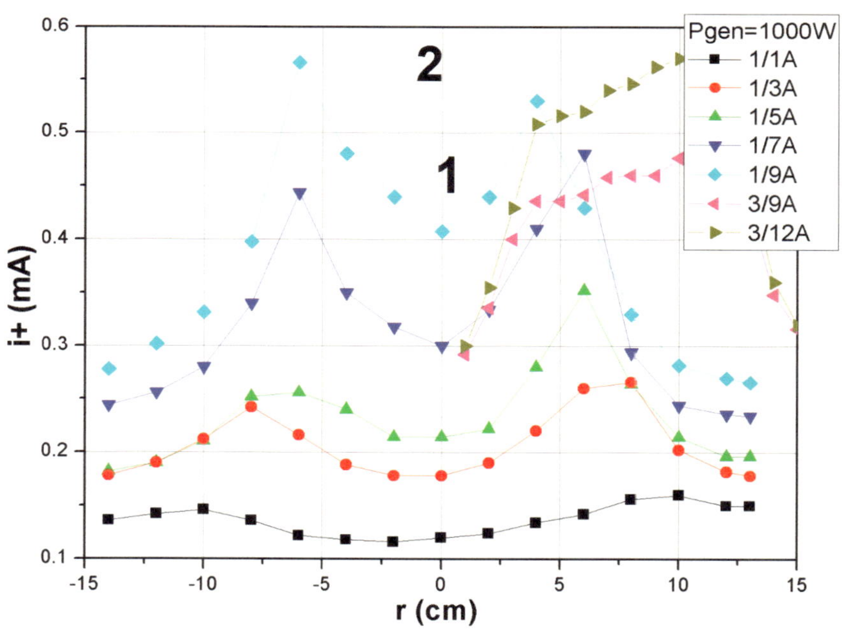

Figure 9.
Radial dependence of the ionic saturation current at different bottom magnet currents while the top magnet current is fixed.

point the RF fields do not yet penetrate the main volume of the plasma. Taking into account that the movement of electrons across the magnetic field is difficult, the distribution of the ionic current with a dip on the axis is observed.

Let us further examine the results of the probe studies in more detail. A typical probe characteristic is shown in **Figure 10**.

As can be seen, the electron energy distribution is close to Maxwell one. **Figure 11** shows the dependence of electron density in the substrate region on the values of currents through the top and bottom magnets at fixed values of current on the top and bottom magnets, respectively.

As can be seen, the range of changes in plasma density corresponds to that which is necessary for the successful implementation of ion assistance technology as part of the magnetron sputtering.

Additional experiments have shown that the best argon pressure range in terms of technological applications is the range between 0.5 and 0.7 Pa. At lower pressures, it is not possible to obtain the required uniformity of the ion current, while an increase in pressure to 1.5 Pa leads to a drop in the values of the ion current.

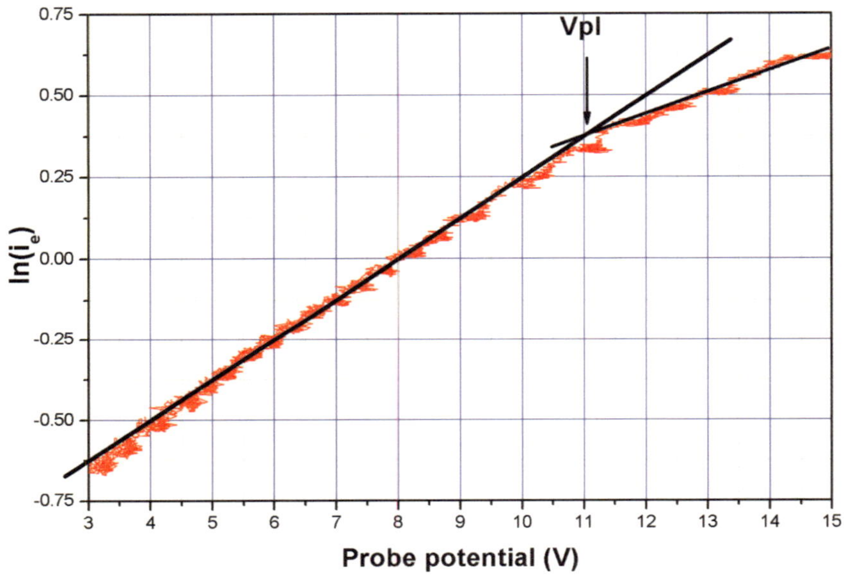

Figure 10.
Dependence of the probe electron current on the probe potential.

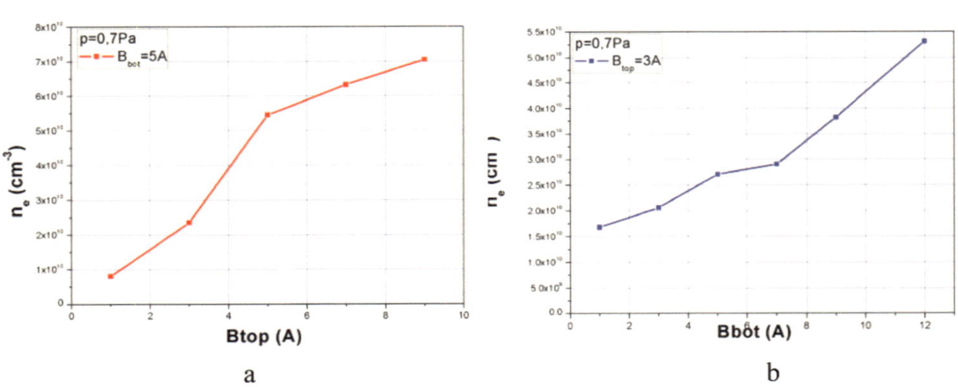

Figure 11.
Dependence of electron density in the substrate region on the values of currents through the top (a) and bottom (b) magnets at fixed values of current on the top and bottom magnets, respectively.

5. Results of the plasma parameter examination when magnetron and gas-discharge sources operate together

At the next stage of the work, the plasma parameters were studied during the simultaneous operation of both the magnetron discharge and inductive RF discharges with external magnetic field. Experiments have shown that the joint operation of magnetron and inductive RF discharges results in a decrease in the threshold pressure, at which magnetron is capable of operating. Titanium spectral lines appeared in the glow spectrum of the magnetron discharge only and at the argon pressure of 0.7 Pa. When magnetron and RF inductive discharges were operating together, the plasma glow spectrum was enriched with titanium lines already at the pressure of 0.3 Pa.

Figure 12 shows the current-voltage characteristics of the magnetron operating at DC mode both independently and together with the gas-discharge source. As can be seen, the sputtering apparatus is capable of operating at substantially lower voltages applied to the cathode.

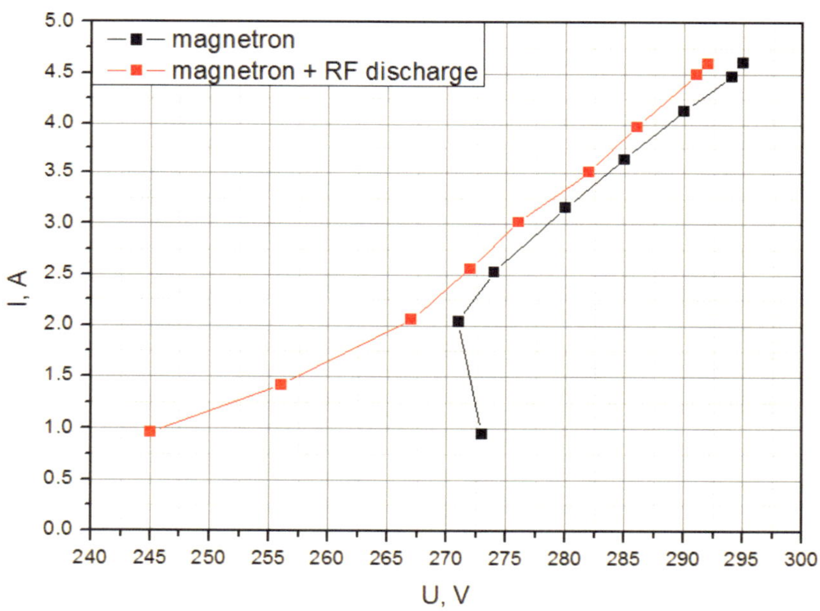

Figure 12.
Current-voltage characteristics of the magnetron operating independently (black curve) and together with the inductive RF discharge with magnetic field (red curve).

6. The effect of assisting ion flow on functional coating structure

To test the effect of ionic stimulation on the properties of functional coatings, thin films were deposited using the magnetron only and using the magnetron and RF sources together with and without applying an additional bias to the substrate. Coatings made of the following materials were used as test samples: Ti, Al, SiAl, SiO_2, and C.

The experiments involving deposition of titanium coatings were aimed at the studying of the effects of the assisting ionic flux magnitude on electro-physical properties of films and their microhardness. It was shown that increase in the power of RF power supply connected to antenna in the range from 0 to 500 W results in almost twofold increase in specific resistivity of the films; at the same time, the microhardness of the films increases by 25%. The observed changes are obviously the result of a change in the film structure. Titanium film morphology studies have shown that the irradiation of films with a flux of accelerated ions leads to a slight decrease in the grain size in the structure of the films. Application of a DC bias to the substrate was accompanied by the smoothing of the surface of films.

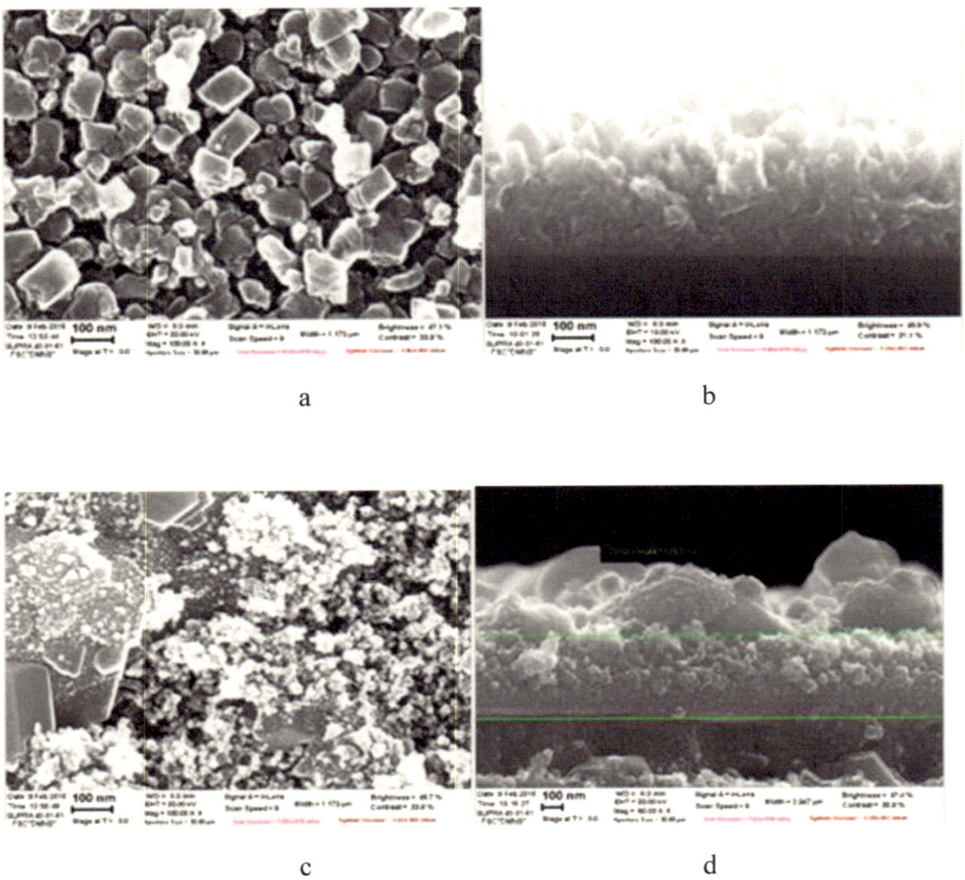

Figure 13.
Aluminum film surface (a and c) and cross-section (b and d).

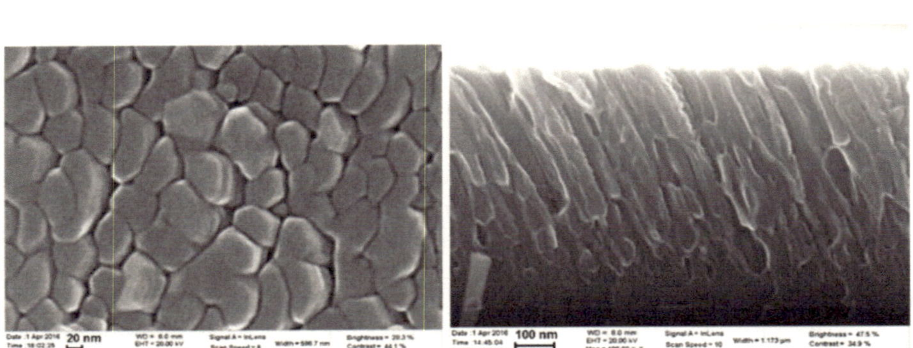

Figure 14.
Film surface and cross-section: Si (Al 10%) deposition, without ionic assistance.

The reduction of grain size in the structure of films in the presence of stimulating ion flux is most clearly seen in the film surface images and the images of aluminum film cleavages, shown in **Figure 13**.

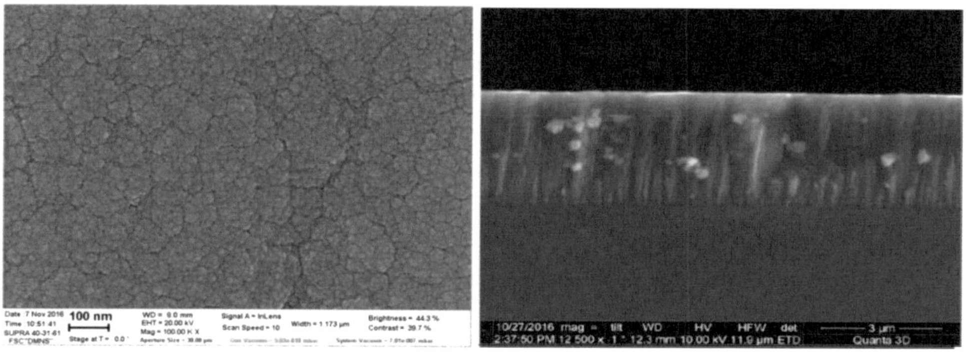

Figure 15.
Film surface and cross-section: Si (Al 10%) magnetron deposition with ionic assistance without applying a bias to substrate.

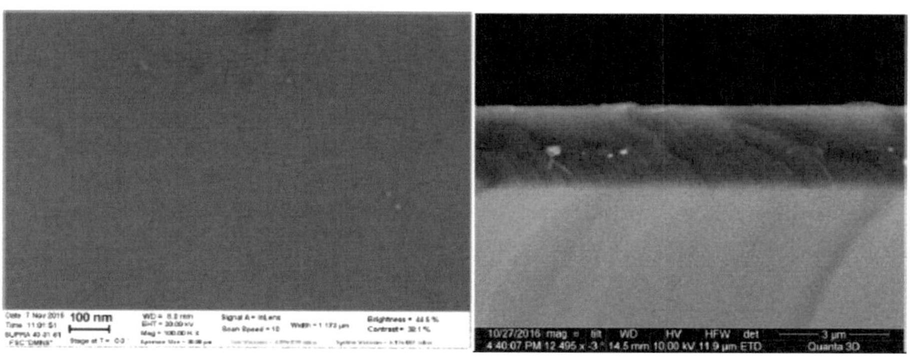

Figure 16.
Film surface and cross-section: Si (Al 10%) magnetron deposition with ionic assistance plus a bias applied to the substrate.

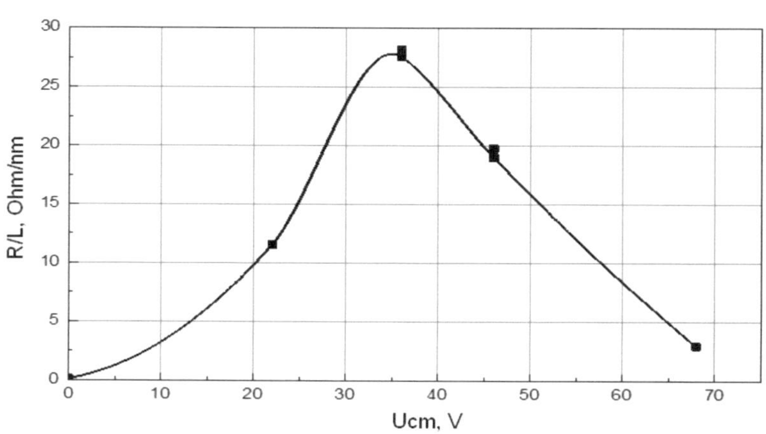

Figure 17.
Dependence of the ratio of film resistance to its thickness on the bias Ucm applied to its substrates.

Interesting results were obtained while depositing silicon with 10% aluminum content. The sample film surface morphology and cleavage are shown in **Figures 14–16**. As can be seen, the samples treated using ionic assistance acquire a columnar structure. When a bias is applied to the substrate, the columnar structure grows at an angle to the normal.

It is known that the ionic stimulation results in significant change in the electrophysical properties of carbon films [28]. **Figure 17** shows the dependence of resistance of a various series of carbon films on the bias voltage [29]. As can be seen, there is a sharp increase in the resistance of the films at the assisting ion energy of 45 eV.

7. Conclusions

The results of the study of plasma parameters in a plasma reactor based on a combination of magnetron and magnetically activated RF discharge indicate the promising outlook as regards to industrial application of the innovative facility. Approaching the region of resonant absorption of RF power by optimizing the magnitude and configuration of the external magnetic field makes it possible to obtain a uniform within 10% radial distribution of the ion current across the diameter of 150 mm. When the RF power supply power is 1000 W, the ion current density on the substrate can be adjusted in the range of 0.1–3 mA/cm^2. The use of ion assisting results in a fundamental change in the structure and properties of functional coatings, deposited using a magnetron.

Acknowledgements

The work was carried out under partial financial support of the RFBR grant 16-02-00646.

Author details

Elena Kralkina[1*], Andrey Alexandrov[1], Polina Nekludova[1], Aleksandr Nikonov[1], Vladimir Pavlov[1], Konstantin Vavilin[1], Vadim Odinokov[2] and Vadim Sologub[2]

1 Physical Faculty of MSU, Moscow, Russian Federation

2 Research Institute of Precision Machine Manufacturing, Zelenograd, Russian Federation

*Address all correspondence to: ekralkina@mail.ru

References

[1] Berghaus B. Improvements in and Relating to the Coating of Articles by Means of Thermally Vaporized Material, U.K. Patent No 510993; 1938

[2] Mattox DM, McDonald JE. Interface formation during thin film deposition. Journal of Applied Physics. 1963;**34**:2493-2494

[3] Mattox DM, Kominiak GJ. Structure modification by ion bombardment during deposition. Journal of Vacuum Science and Technology. 1972;**9**:528-532

[4] Colligon JS. Ion-assisted sputter deposition. Philosophical Transactions of the Royal Society A. 2004;**362**:103-116

[5] Mattox D. The Foundation of Vacuum Coating Technology. N.Y.: Noyes Publications; 2003.534 p

[6] Zhurin VV. Industrial Ion Sources. Broad Beam Gridless Ion Source Technology. N.Y.: Wiley-VCH Verlag & Co; 2012. 382 p

[7] Harper JME, Cuomo JJ, Kaufman HR. Technology and applications of broad beam ion sources used in sputtering. Part II. Applications. Journal of Vacuum Science and Technology. 1982;**21**:737-756

[8] Kaufman HR, Harper JME. Ion assists applications of broad beam ion sources. Proceedings of SPIE. 2004;**5527**:50-68

[9] Petrov IN, Adibi F, Greene JE, Hultman L, Sundgren JE. Average energy deposited per atom: A universal parameter for describing ion-assisted film growth. Applied Physics Letters. 1993;**63**:36-38

[10] Winters H et al. Influence of energy reflected from the target on thin film characteristics. Journal of Vacuum Science and Technology. 1998;**11**:657-663

[11] Rossnagel S. Thin film deposition with physical vapor deposition and related technologies. Journal of Vacuum Science and Technology. 2003;**21**:74-87

[12] Fulton M. Application of ion assisted deposition using gridless end-Hall ion source for volume manufacturing of thin film optical filters. In: Proceedings of Optical Interference Coatings; 1994. p. 2253

[13] Kaufman HR, Robinson RS. Operation of Broad-Beam Sources. Alexandria, Virginia: Commonwealth Scientific Corporation; 1987. 453 p

[14] Shpak A, Rudenko E, Korotash I, Senyuk V, Shamrai K, Odinokov V, et al. Plasma source of low temperature formation of the nanoclusters of metal–catalyst. Nanoindustry (Russian Federation). 2009;**4**:12-15

[15] Aleksandrov AF, Vavilin KV, Kral'kina EA, Neklyudova PA, Nikonov AM, Pavlov VB, et al. Plasma parameters in the reactor with simultaneous magnetron discharge and inductive radio-frequency discharge in the presence of external magnetic field. Journal of Communications Technology and Electronics. 2018;**63**:374-380. DOI: 10.22184/1993-8578.2016.70.8.104.108

[16] Aleksandrov AF, Petrov AK, Vavilin KV, Kral'kina EA, Neklyudova PA, Nikonov AM, et al. Influence of external conditions on physical processes and plasma parameters in a model of a high-frequency hybrid plasma system. Russian MicroElectronics. 2016;**45**:433-441

[17] Aleksandrov AF, Petrov AK, Vavilin KV, Kralkina EA, Neklyudova PA, Nikonov AM, et al. Investigation of the helicon discharge plasma parameters

in a hybrid RF plasma system. Plasma Physics Reports. 2016;**42**:290-292. DOI: 10.1132/S1063780X16030016

[18] Vavilin KV, Kralkina EA, Nekludova PA, Petrov AK, Nikonov AM, Pavlov VB, et al. Hybrid plasma system for magnetron deposition of coatings with ion assistance. Journal of Physics Conference Series. 2015;**669**:1-4. DOI: 10.1088/1742-6596/669/1/012033

[19] Johnson WL. Electrostatically-shielded inductively-coupled RF plasma source. In: Popov OA, editor. High Density Plasma Sources. NY: Noyes Publications; 1996. pp. 100-114

[20] Chen FF. Helicon plasma sources. In: Popov OA, editor. High Density Plasma Sources; 1996. pp. 1-75

[21] Shinohara S. Helicon high density plasma sources: Physics and applications. Advances in Physics. 2018;**3**(1):1420424. DOI: 10.1080/23746149.2017.1420424

[22] Chen F. The low-field density peak in helicon discharges. Physics of Plasmas. 2003;**10**:2586-2592. DOI: 10.1063/1.1575755

[23] Chen FF, Jiang X, Evans JD, Tynan G, Arnush D. Low-field helicon discharges. Plasma Physics and Controlled Fusion. 1997;**39**:A411-A420

[24] Kral'kina EA. Low-pressure radio-frequency inductive discharge and possibilities of optimizing inductive plasma sources. Physics-Uspekhi. 2008;**51**:493-512. DOI: 10.1070/PU2008v051n05ABEH006422

[25] Kralkina EA, Rukhadze AA, Nekliudova PA, Pavlov VB, Petrov AK, Vavilin KV. RF power absorption by plasma of low pressure low power inductive discharge located in the external magnetic field. AIP Advances. 2018;**8**:035217. DOI: 10.1063/1.5023631

[26] Chen FF, Jiang X, Evans JD. Plasma injection with helicon sources. Journal of Vacuum Science and Technology A. 2000;**18**:2108-2115. DOI: 10.1116/1.1289537

[27] Petrov AK. Physical processes in the two-chamber inductive RF plasma source with the external magnetic field [thesis]. Moscow: Physical faculty of Moscow State University; 2016 [in Russian]

[28] Guseva MB, Savchenko NF, Babaev VG. The influence of ion irradiation on the structure and properties of amorphous carbon films. Radiation Effects Letters. 1986;**87**:215224

[29] Grushin M, Kralkina E, Nekludova P, Nikonov A. Influence of the ion assistance on the physical properties of the carbon films obtained by the magnetron sputtering. In: Proceedings of the Conference "Low Temperature Plasma in the Processes of Functional Coatings Deposition". Kazan; 2018 [in Russian]

Section 2

Modelling and Simulation

Chapter 6

Study on Specific Coefficient in Micromachining Process

Sung-Hua Wu

Abstract

The study proposed an important micro-specific coefficient based on the mathematical modeling of micro-cutting resistance to predict the mechanic conditions at cutter-edge radius. For the steady-state chip formation in the micro-cutting process, the differential angle is usually constant, and the plowing angle and rake angle are relative to the tool-edge radius, cutting resultant force, plowing resistance, surface roughness, and shearing resistance on the tool-workpiece. The optimal process included a cut of depth of 0.001 mm, cutting length of 0.003 mm, cutter-edge temperature of 38°C, and an edge radius of 0.0005 mm on workpiece Al-7075; the optimal cutting force in x-axis was 0.0005 N (Avg.) and the optimal cutting force in y-axis was 0.00028 N (Avg.) for better surface roughness Ra = 0.16. The higher temperature was 42.16°C on the workpiece and tool HSS, and the maximum strain rate occurred on the chip shearing zone was 9.33E6 (/s), which obeyed the generalized cutting criterion by numerical analysis. While the micro-specific coefficient is close to 1, the plowing zone will increase friction, stress, resistance, and even cutting excited vibration, resulting in discontinuous chipping. Besides, the process developed the micro-MDOF cutting dynamics model and applied a fractal equation to simulate the micro-cutting process. The validation can be proved as the derived theory agreed well with the simulation in the micro-cutting process.

Keywords: specific coefficient, fractal equation, micro-cutting, steady-state chip formation, micro-MDOF cutting dynamics

1. Introduction

Micro-cutting process for depth of cut and feed is so small that it is very hard to observe under a microscope. Due to the size effect of the cutter-edge radius for larger influence on plowing and shearing zones, it results in increased friction heat, cutting force and, furthermore, specific energy. The study proposed the micro-resistance model of plowing and shearing in the quasi-state cutting process by analytic geometry. Using the fractal equation can help us to build the relations of chip fractal geometry and specific energy. The famous scholar Mandelbrot gave the set [1] is a compact set (**Figure 1**), where resisted of complex numbers c, and he studied space of quadratic polynomials and proposed the function $f_c\ (z)$ existed non-diverged and iterated from z equals zero. The fractal properties have self-similarity, hyperbolic components, and local connectivity from a topological space x. The author observed the properties of self-similarity of fractal existing on undeformed chip geometry through a series of cross-section projections on a

IntechOpen

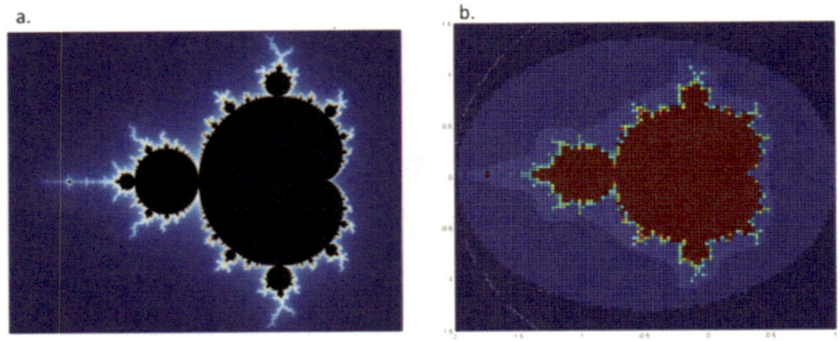

Figure 1.
(a) Mandelbrot set of fractal; and (b) the author's program by complex [5].

sinusoidal wave. The undeformed chip geometry through a series of cross-section projection on a sinusoidal wave by mathematical modeling belongs to a kind of Koch curve transformation, where the Koch curve belongs to a kind of triangle fractal as shown in **Figure 2(d)**. The fractal similarity can be applied in microanalysis of chip formation through enlarging the contour of the chip topograph as shown in **Figure 2(c)**. From another view as shown in **Figure 2(f)**, the fractal geometry is similar to a series of projections of sinusoidal multi-cutter as a rectangle (**Figure 3(d)**). Through analytic geometry, the fractal differential equation can be established to understand chip shapes, chip load, tool geometry, and specific energy. From the view under a microscope, the chip formation is similar (**Figure 2 (d)–(f)**) in different views of fractal geometry. Hence, the fractal topograph of **Figure 2** is very important to establish a generalized mathematical model by the analysis of chip fractal geometry. By Fractal mathematical method, the calculation can be achieved less hardware and higher precisely results as **Figure 1(b)**. Due to the excellent calculation and algorithm, material processing or cutting dynamics can

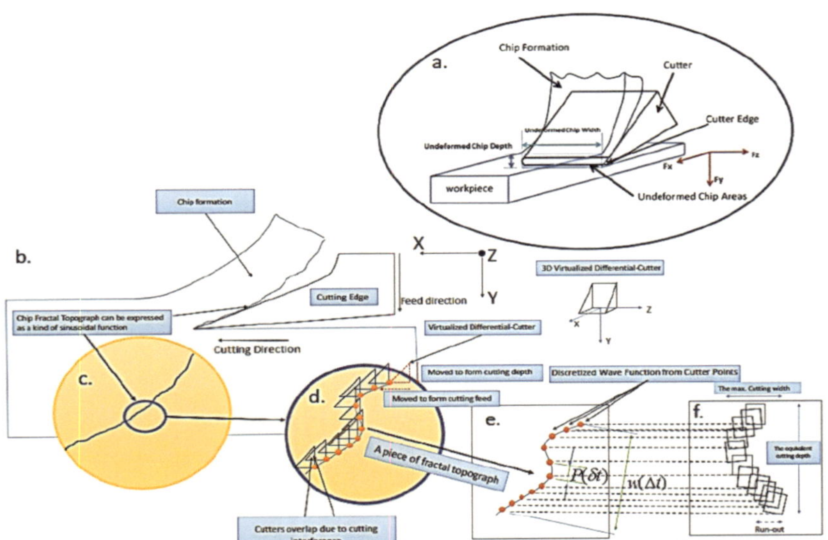

Figure 2.
Undeformed chip areas and deformed chip shapes: (a) the 3-D expression of undeformed cutting areas and chip formation; (b) the side view for chip fractal topograph; (c) to take a part of the contour along the chip edge; (d) the topograph can be simulated by a lot of differential-cutters moved to form cutting depth and feed; (e) the side view of chip topograph can be resisted of cutters under cutting interference in differential time; (f) the state-space transferring for chip fractal topograph into undeformed cutting areas (fractal self-similarity) from (e) to (f).

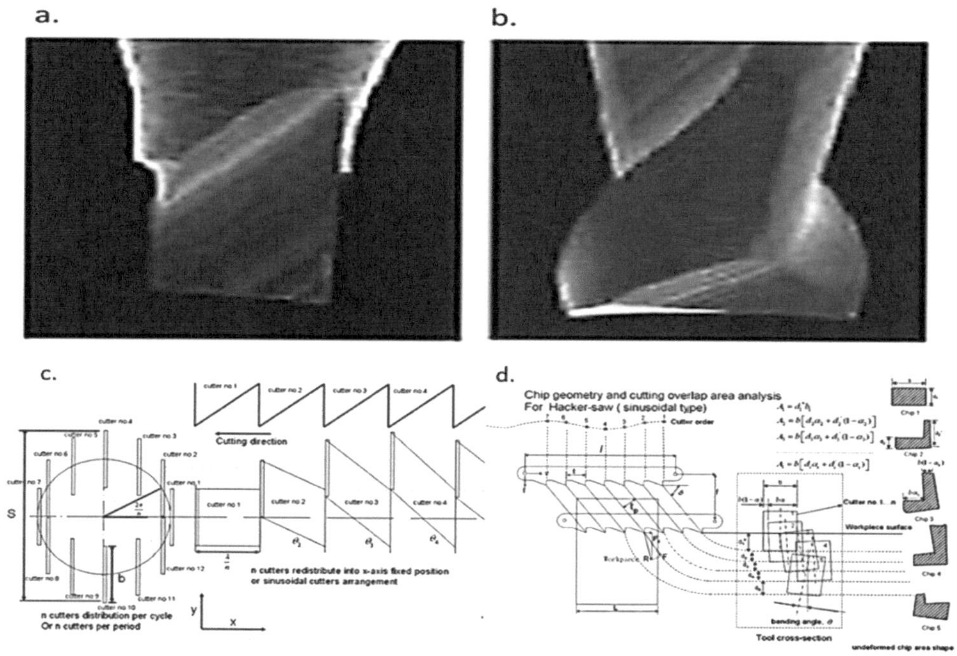

Figure 3.
General micro-cutter: (a) single cutter—r = 10–50 μm; (b) double cutter—helical angle 30° and r = 30 – 50 μm; (c) unit cutters of generalized sinusoidal multi-cutters for the analytic coordinate; and (d) analysis on chip fractal self-similarity of sinusoidal cutter order.

be analyzed by specific energy obtained from undeformed chip formation (**Figures 2** and **3**). Besides, the theoretical modeling should consider the surface condition of workpiece because the surface roughness and tool size are enlarged in micro-dimension. So, the surface conditions need to be expressed as shown in **Figure 4**. Hence, the micro-MDOF cutting dynamics model can be developed as Eqs. (22)–(32).

1.1 Paper review

Cao et al. [2] found that the cutting edge radius affects the microscale in the cutting process at a smaller uncut chip thickness by altering the effective rake angle, enhancing the plowing effect and the work affecting the material deformation process, expanding and widening the plastic deformation zone, and causing higher energy dissipation due to increased tool-chip contact length. The author wants to expand the work to develop the theoretical model by advanced mathematics. On the other hand, the study of Cao et al. [2] unestablished the micro-specific coefficient to investigate the relations of factors on plowing, shearing, cutter-edge radius, cutting heat, and stain rate. To extend the study of Cao et al. [2], the fractal equation proposed in Wu [3–5] can be applied on the analysis of variations in chip shapes and undeformed chip formation areas of specific energy obtained. Wu [5] considered the vertical force F_x and horizontal force F_y existing in the mathematical relations as the objective function of the specific energy constant that can achieve the maximum material rate (MMR) through optimization, while a single cutter achieved a steady-state chip formation. The definition of the proportion Kr equals Fy/Fx and the optimal Kr = 0.602 for tool HSS and workpiece Al-7075 if the maximum material rate is achieved. The traditional Kr obtained was assumed as the specific energy constant and known cutting force without considering the microscope size effect.

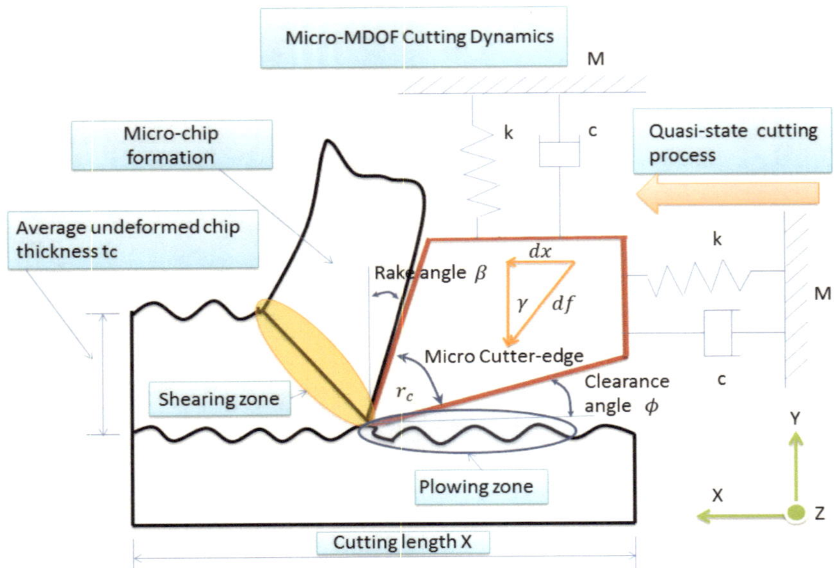

Figure 4.
Micro-MDOF modeling demonstration.

Otherwise, the specific energy and cutting force need to be rebuilt if the material property of workpiece-tool changed. The strategy of micromachining for MMR is also dependent on the optimal specific coefficient obtained by the process, achieving the steady-state chip formation under the known specific energy. Another result expressed a larger excited-vibration while Kr was close to 1 (about 1.1–1.3) through two cutters' simulated design. So, the Kr should be an important factor affecting the cutting stability and MMR in the steady-state cutting process. From the observation of experimental results, a few cutters showed cutter-edge-crack wear similar to the discontinuous wear, and others showed flank wear similar to the continuous wear. The results reminded author Wu that there were some reasons to explain how the shearing and plowing effect occurred on cutter edge in the micromachining process. The study [5] can be extended to investigate the micro-specific coefficient through mathematical derivation. About micromachining, micro-cutting is a general process for the material removal. The key point of micromachining is how to achieve better production precision and reasonable error, even more/less wear on the tool and the maximum material removal rate. David et al. [6] proposed a technique to fabricate the micro-tool, which can achieve a 25 μm diameter and a cutting edge radius curvature of less than 0.1 μm in 6061-T4 aluminum, 4340 steel, and polymethyl methacrylate by focused ion beam. Dib et al. [7] studied minimum chip thickness determination by means of cutting force signal in micromachining. An aluminum alloy (RSA6061-T6) was cut by micro-end mills of carbide. Due to the plowing effect, the material may achieve a height 2–4 times the minimum thickness by using the analysis of cutting force signal of a mini dynamometer. Chae et al. [8] investigated the application by micro-cutting, where a process is expressed to use micro-meso-scale fabrication by miniaturized mechanical material removal processes in creating 3D components using a variety of engineering materials. One of the biggest challenges is to maintain cutting force under a critical limit tool wear and improve productivity. The size effect is also a key question that needs to consider micro-tool geometry, minimum chip thickness, and process parameters. Oliveira et al. [9] studied the relations between size effect and minimum chip thickness in the micro-milling process. The results showed that the minimum uncut chip thickness varies practically from 1/4 to 1/3 of tool cutting edge. Zhang et al. [10] developed the cutting force in the micro-end-milling process. By considering the theoretical instantaneous uncut chip thickness, the certain critical chip thickness value and minimum uncut chip thickness can be derived by material removal mechanisms. For the micromachining process, tool wear is another key question, where the influence mainly comes from the plowing effect causing larger heat. These two key questions need a good criterion or equation to explain the compliant process. The author Wu expects to find out the key factors for better production quality and establish the equation through the study. Tansel et al. [11, 12] applied a tool wear estimation model to predict the cutting force in micromachining by a neural-network-based method, where the final results achieved to the maximum and minimum errors of 8.84 and 5.09%, respectively. Malekian et al. [13] proposed the minimum uncut chip thickness (MUCT) to study the tool-edge radius in micromachining, resulting in increased machining forces that affect the surface integrity of the workpiece. The results presented that the minimum uncut chip thickness (MUCT) was based on the minimum energy principle and the infinite shear strain and the stagnation point with respect to the edge radius. Wu et al. [14] proposed a groove-sawing model on sinusoidal multi-cutters to investigate the relations between specific energy distribution and the location of tool wear. Hence, the cutters' arrangement, cutters' material, uncut chip shapes, and specific energy can be defined by a fractal equation. Process stability is worthy to understand the development, which can be applied in machining, manufacturing, and design. Afazov et al. [15] found the increase of the cutting forces associated with the variation of the friction conditions between the tool and workpiece contact and stability lobes obtained for different edge radii and rake angles of 0 and 8° in the micro-milling process. The results caused the stability of the process obtained for a wide range of cutter-edge radii, feed rates, and run-out lengths. Filiz et al. [16, 17] have analytically investigated according to a three-dimensional model for micro-end-milling dynamics. The modeling included cross-section and fluted (pre-twisted) geometry and bending and coupled axial/torsional vibrations. Afazov et al. [18, 19] chattered on modeling in micro-milling, where the theory was based on the two degrees of freedom of dynamic parameters for the tool-holder-spindle assembly and the micro-milling cutting forces. Guo et al. [20] applied cutting force modeling for nonuniform helix tools and chip thickness in a five-axis flank milling process by an instantaneous uncut chip thickness (IUCT) model. The final results showed that the difference between helix angles and two edges is small, and the little effect occurred on the cutting forces for the cutting force coefficient prediction error. Malekian et al. [21] established the plowing force model, which can take the effect of elastic recovery based on the interference volume between the tool-workpiece, where the mechanistic model has been verified with experimental cutting force measurements of Al-6061.

2. Theoretical modeling

The process developed the micro-MDOF cutting dynamics model and micro-fractal equation to simulate the micro-cutting process. Through fractal mathematics, the results presented optimal geometric parameters for micro-cutting simulation for tool HSS and workpiece Al-7075 as shown in **Figures 4–6**. The study found that the micro-specific coefficient $\hat{K}_t$ applied in the micromachining process is to explain the influence of shearing and plowing at cutter-edge radius.

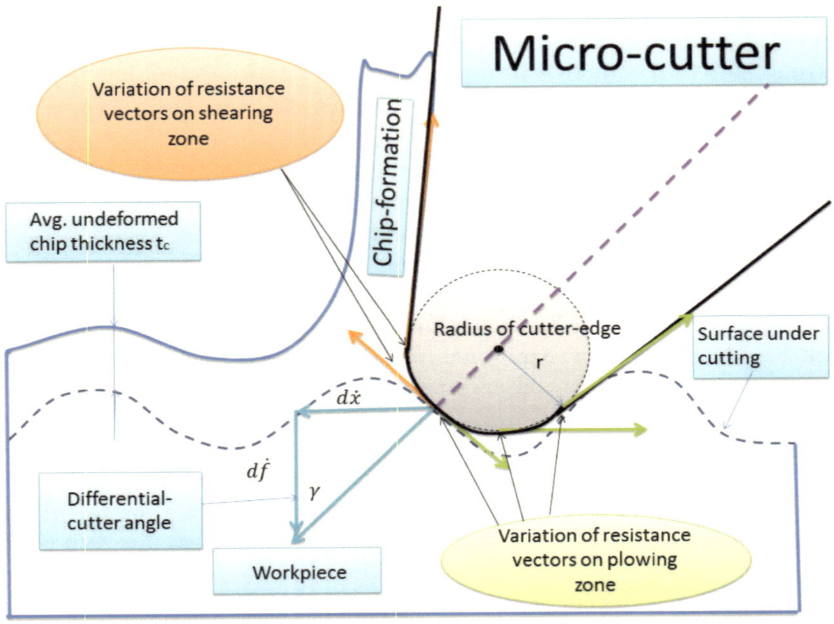

Figure 5.
The relations between micro-cutter-edge radius and chip formation.

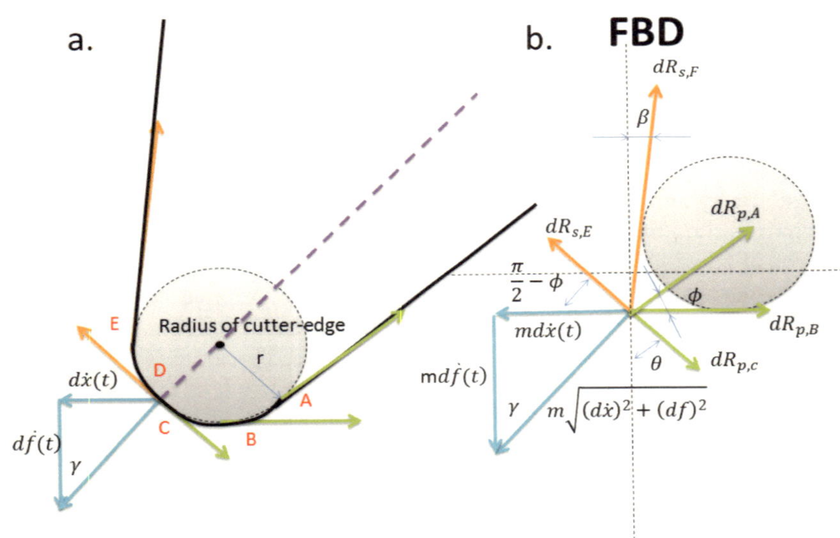

Figure 6.
(a) Vectors of resistance at cutter edge; (b) FBD in the quasi-state micro-cutting process.

2.1 Establishment of the specific coefficient for plowing and shearing resistance in the quasi-state micro-cutting process

The quasi-state micro-cutting process is a dynamical balance process for cutting resistance and cutting force. The three points C, D, and E are aligned as the same points from **Figure 6(a)**. Firstly, the free body diagram (FBD) for plowing and shearing resistance in the quasi-state micro-cutting process should be established as Eqs. (1) and (2):

$$x \leftarrow + : md\dot{x}(t) + dR_{s,E}\cos\left(\frac{\pi}{2}-\phi\right) - dR_{s,F}\sin(\beta) - dR_{p,A}\cos(\phi) - dR_{p,B} \\ -dR_{p,C}\sin(\theta) = 0 \tag{1}$$

$$y \downarrow + : md\dot{f}(t) - dR_{s,E}\sin\left(\frac{\pi}{2}-\phi\right) - dR_{s,F}\cos(\beta) - dR_{p,A}\sin(\phi) \\ +dR_{p,C}\cos(\theta) = 0 \tag{2}$$

Thus, Eqs. (3) and (4) can be obtained simultaneously as follows:

$$\Rightarrow md\dot{x}(t) + \cos(\phi)\left[dR_{s,E} - dR_{p,A}\right] - dR_{s,F}\sin(\beta) - dR_{p,C}\sin(\theta) = dR_{p,B} \tag{3}$$

$$\Rightarrow md\dot{f}(t) - \sin(\phi)\left[dR_{s,E} + dR_{p,A}\right] - dR_{s,F}\cos(\beta) + dR_{p,C}\cos(\theta) = 0 \tag{4}$$

For the orthogonal cutting process,

$$\because dR_{s,E} \perp dR_{p,A} \\ \therefore md\dot{x}(t) - dR_{s,F}\sin(\beta) - dR_{p,C}\sin(\theta) = dR_{p,B} \\ \Rightarrow md\dot{x}(t) = dR_{s,F}\sin(\beta) + dR_{p,C}\sin(\theta) - dR_{p,B} \tag{5}$$

Then,

$$md\dot{f}(t) - dR_{s,F}\cos(\beta) + dR_{p,C}\cos(\theta) = 0 \tag{6}$$

Hence,

$$\Rightarrow \tan(\gamma) = \frac{d\dot{x}(t)}{d\dot{f}(t)} = \frac{dR_{s,F}\sin(\beta) + dR_{p,C}\sin(\theta) - dR_{p,B}}{dR_{s,F}\cos(\beta) - dR_{p,C}\cos(\theta)} \tag{7}$$

Due to the size effect with considering the cutter-edge radius r in micromachining process, the differential-cutter angle γ is relative to two factors: plowing angle θ and rake angle β. For the steady-state chip formation in the micro-cutting process, the rake angle β is usually a constant, but the plowing angle θ is relative to the tool-edge precision and surface roughness on the workpiece. For the condition of resistance in the micro-cutting process, cutting resistance on points F, B, and C has a large influence on the steady-state chip formation process. Arc $\widehat{FBC}$ of cutter-edge radius r about $\frac{2\pi}{3}r$ is an important factor, which is needed to consider tool stiffness, geometric design, and material properties in the machining process.

While $\frac{d\dot{x}(t)}{d\dot{f}(t)} = 1$ or close to 1, called plowing and shearing coupling, large plowing resistance $R_{p,B}$ and vibration on the workpiece surface will occur because the plowing zone at B point is very obvious from Eq. (8). The condition should cause an increase in surface roughness:

$$\frac{d\dot{x}(t)}{d\dot{f}(t)} = \hat{\mathrm{K}}_{\mathrm{r}} = \frac{dR_{s,F}\sin(\beta) + dR_{p,C}\sin(\theta) - dR_{p,B}}{dR_{s,F}\cos(\beta) - dR_{p,C}\cos(\theta)} = 1 \tag{8}$$

where $\hat{\mathrm{K}}_{\mathrm{r}}$ means the specific coefficient of micro-cutting.

Furthermore,

$$\Rightarrow dR_{s,F}\sin(\beta) + dR_{p,C}\sin(\theta) - dR_{p,B} = dR_{s,F}\cos(\beta) - dR_{p,C}\cos(\theta) \tag{9}$$

Going on,

$$\Rightarrow dR_{s,F}[\sin(\beta) - \cos(\beta)] + dR_{p,C}[\sin(\theta) + \cos(\theta)] = dR_{p,B} \tag{10}$$

From integration, we obtain

$$\Rightarrow \int dR_{p,B} = \int [\sin(\beta) - \cos(\beta)]dR_{s,F} + \int [\sin(\theta) + \cos(\theta)]dR_{p,C} \tag{11}$$

To obtain the function of the plowing at B in micromachining process,

$$\Rightarrow R_{p,B} = \int [\sin(\beta) - \cos(\beta)]dR_{s,F} + \int [\sin(\theta) + \cos(\theta)]dR_{p,C} + C \tag{12}$$

Equation (12) is an important result for plowing and shearing influence of micro-cutting under the size effect. While $\frac{d\dot{x}(t)}{d\dot{f}(t)} = \hat{K}_r$ is close to 1, the plowing zone will increase friction, stress, resistance, and even cutting excited-vibration or chipping, resulting in discontinuous chipping. The way to improve is to design the rake angle and cutter-edge radius or raise the tool precision and material.

Although $R_{p,B} = R_{p,B}(R_{s,F}, R_{p,C}, \beta, \theta)$ denoted the resistance function in the micromachining process, the function can be expanded to investigate the relations of factors $R_{p,B}$, $R_{s,F}$, $R_{p,C}$, β, θ combining micro-cutting force, shear stress, strain, and strain rate.

From the view of the micro-cutting process, the definition of micro-specific coefficient is $\hat{K}_r = \frac{d\dot{x}}{d\dot{f}}$, and the definition of average micro-specific coefficient is $\hat{K}_{r,avg} = \frac{d\dot{x}_2 - d\dot{x}_1}{d\dot{f}_2 - d\dot{f}_1} = slope$; if $\hat{K}_r = 0$, Eq. (8) can be rewritten as follows:

$$\begin{aligned} & dR_{s,F}\sin(\beta) + dR_{p,C}\sin(\theta) - dR_{p,B} = 0 \\ & \Rightarrow dR_{p,B} = dR_{s,F}\sin(\beta) + dR_{p,C}\sin(\theta) \\ & \Rightarrow R_{p,B} = \int \sin(\beta)dR_{s,F} + \int \sin(\theta)dR_{p,C} + C \end{aligned} \tag{13}$$

The result still expresses that the function $R_{p,B} = R_{p,B}(R_{s,F}, R_{p,C}, \beta, \theta)$ existed.

2.2. Generalized chip load of micro-cutter by the fractal differential equation

According to sinusoidal multi-cutters, the cutters' interference ratio of cutter arrangement can be written as Eq. (14):

$$\alpha_{n+1} = 1 - \frac{\lambda}{nb}(\tan\vartheta_{n+1} - \tan\vartheta_n) \tag{14}$$

where α_{n+1} means $(n+1)^{th}$ cutter, λ denotes the length of unit wave, n means cutter number of period, b means cutter width, ϑ_{n+1} and θ_{n+1} means the banding angle of the unit cutter. The equivalent cutting depth of the cutting interference can be presented as Eq. (15):

$$d_{i,eq}=d_1^*\left(\frac{\alpha_i}{1-\alpha_i}\right)^{n-1}+\left(d_{i-1}-d_{i-2}\right)\left(1-\alpha_i\right)+d_1^* \tag{15}$$

where d_1^* means the initial cutting depth and d_i means the transient cutting depth. Considering the cutting depth and cutting width, Eq. (16) can be obtained if the frictional energy dissipated:

$$\begin{aligned}P_i &= A_i = b_i d_i\\ &=\left(1-\alpha_i\right)b^*\left[d_1^*\left(\frac{\alpha_i}{1-\alpha_i}\right)^{n-1}+\left(d_{i-1}-d_{i-2}\right)\left(1-\alpha_i\right)+d_1\right]\end{aligned} \tag{16}$$

A_i means the undeformed chip shapes, which is also similar to chip fractal geometry areas P_i in the steady-state chip formation process:

$$\begin{aligned}&P\left(d_i,\alpha_i,b,n,\theta_i\right)\\ &=\left(1-\alpha_i\right)b^*\left[d_1^*\left(\frac{\alpha_i}{1-\alpha_i}\right)^{n-1}+\left(1-\alpha_i\right)\left(d_{i-1}-d_{i-2}\right)+d_1^*\right]\end{aligned} \tag{17}$$

Simultaneously,

$$\tan\vartheta = R\sin\left(i\times\frac{2\pi}{n}\right)/t_p \tag{18}$$

where R means amplitude of sinusoidal-set cutters, t_p means cutter pitch, and ϑ denotes the bending angle of sinusoidal multi-cutters.

Equation (19) will delete the overlap ratio $\alpha_i = 0$ if the sinusoidal multi-cutters become a single cutter. On the other hand,

$$\begin{aligned}P &= (1-\not{\alpha}_i)b^*\left[{d_1}^*\left(\frac{\not{\alpha}_i}{1-\not{\alpha}_i}\right)^{n-1}+(1-\not{\alpha}_i)(d_{i-1}-d_{i-2})+{d_1}^*\right]\\ &= b^*{d_1}^*\end{aligned} \tag{19}$$

where P means the generalized chip load of the micro-cutter by the fractal differential equation; the average cutting depth d_1^* can be obtained through a few experiments and simulation or specific energy $\mathrm{K_t}$ can be obtained if the cutting length b^* is usually constant in the micro-cutting process. Average specific energy $\mathrm{K_t}$ can be obtained through unit cutting areas of multi-cutters by average chip load of each chip formation as Eq. (20):

$$\overline{\mathrm{K}}_{\mathrm{t}} = \frac{W}{\Delta V} = \frac{P\not{L}}{b^*{d_1}^*\not{L}} = \frac{P}{b^*{d_1}^*} \tag{20}$$

The cross-sections Type A, Type B, and Type C of undeformed chip shapes can be derived as Eq. (21) and **Figure 7**:

$$\begin{aligned}P_{typeA} &= \overline{\mathrm{K}}_{\mathrm{t}}b^*{d_1}^*\ \textit{for Type A}\\ P_{typeB} &= \overline{\mathrm{K}}_{\mathrm{t}}b^*(1-\tan\varphi){d_1}^*\ \textit{for Type B}\\ P_{typeC} &= \frac{1}{2}\overline{\mathrm{K}}_{\mathrm{t}}b^*{d_1}^*\ \textit{for Type C}\end{aligned} \tag{21}$$

2.3 Micro-MDOF cutting dynamics equation

A tiny cutter should have a tiny mass δm, damping c, and elastic constant k with micro-dynamic loading P as shown in **Figure 4**. It needs to construct a dynamic equation as Eq. (22):

$$[\delta m]\{\ddot{x}\} + [c]\{\dot{x}\} + [k]\{x\} = \{p\} \tag{22}$$

Assume the initial condition as

$$\begin{aligned} \{\overline{x}(0)\} &= [\Phi]\{q(0)\} \\ \{\dot{\overline{x}}(0)\} &= [\Phi]\{\dot{q}(0)\} \end{aligned} \tag{23}$$

Substitute $\{x_i\}^T$ into Eq. (22),

$$[\delta m]\{\ddot{x}\} + [c]\{\dot{x}\} + [k]\{x\} = \{p\} \tag{24}$$

Then,

$$\begin{aligned} &\{\tilde{x}_i\}^T[\delta m][\Phi]\{\ddot{q}\} + \{\tilde{x}_i\}^T[c][\Phi]\{\dot{q}\} + \{\tilde{x}_i\}^T[k][\Phi]\{q\} = \{\tilde{x}_i\}^T\{p\} \\ &(i = 1, 2, 3\ldots, n) \end{aligned} \tag{25}$$

For orthogonal conditions,

$$\begin{aligned} \{\tilde{x}_j\}^T[\delta m]\{\tilde{x}_i\} &= 0, (i \neq j) \\ \{\tilde{x}_j\}^T[c]\{\tilde{x}_i\} &= 0, (i \neq j) \\ \{\tilde{x}_j\}^T[k]\{\tilde{x}_i\} &= 0, (i \neq j) \end{aligned} \tag{26}$$

Going on,

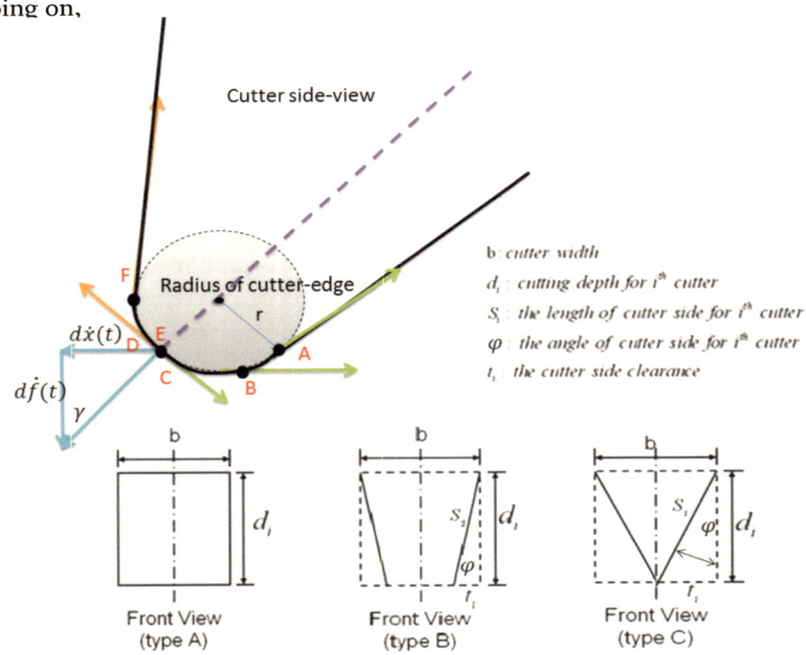

Figure 7.
Variation of cross-section shapes for undeformed chip formation from Type A to Type C.

$$\delta m_i \ddot{q}_i + c_i \dot{q}_i + k_i q_i = p_i, (i = 1, 2, 3, ..., n) \tag{27}$$

To define

$$\begin{aligned} \delta m_i &= \{\tilde{x}_i\}^T [\delta m] \{\tilde{x}_i\} \\ c_i &= \{\tilde{x}_i\}^T [c] \{\tilde{x}_i\} \\ k_i &= \{\tilde{x}_i\}^T [k] \{\tilde{x}_i\} \\ p_i &= \{\tilde{x}_i\}^T \{p\} \end{aligned} \tag{28}$$

Hence,

$$\ddot{q}_i + \frac{c_i}{\delta m_i} \dot{q}_i + \frac{k_i}{\delta m_i} q_i = \frac{p_i}{\delta m_i} \tag{29}$$

Assume

$$\begin{aligned} \frac{c_i}{\delta m_i} &= 2\xi_i \omega_i \\ \frac{k_i}{\delta m_i} &= {\omega_i}^2 \end{aligned} \tag{30}$$

Furthermore,

$$\ddot{q}_i + 2\xi_i \omega_i \dot{q}_i + {\omega_i}^2 q_i = \frac{p_i}{\delta m_i} \tag{31}$$

To solve, the micro-MDOF cutting dynamics equation can be obtained as Eqs. (31) and (32):

$$\begin{aligned} q_i = q_i(t) = e^{-\xi_i \omega_i t} \left[\frac{\dot{q}_i(0) + \xi_i \omega_i q_i(0)}{\omega_p} \sin(\omega_p t) + q_i(0) \cos(\omega_p t) \right] + \cdots \\ (i = 1, 2, 3, ..., n) \end{aligned} \tag{32}$$

where $\omega_p = \omega_i \sqrt{1 - {\xi_i}^2}$.

Equation (32) can be expanded by Taylor series expansion or numerical analysis.

In order to obtain the temperature variation of the cutter edge and chip surface, where the Johnson-Cook equation is used as Eq. (33), the temperature conduction between tool and workpiece is assumed. The study does not consider the cutting temperature conducting into the air:

$$\sigma_{eq} = (A + B\varepsilon_n) \left(1 + \mathrm{C} \ln \left(\frac{\dot{\varepsilon}}{\dot{\varepsilon}_0}\right)\right) \left(1 - \left(\frac{T - T_{room}}{T_m - T_{room}}\right)^m\right) \tag{33}$$

where T_m means the melting point temperature, T_{room} means the environmental temperature, T means the workpiece temperature, A means the yielding stress, B means the strain factor, n means the strain coefficient, m means the temperature coefficient, $\dot{\varepsilon}$ means the plastic strain ratio, and ε_0 means the stain ratio. The simulation processed the orthogonal micro-cutting by the Lagrangian finite element method and numerical analysis. Eq. (33) offers the cutting temperature distribution on the tool-workpiece.

3. Results and discussion

The study by micro-MDOF cutting dynamics simulated micro-cutting process according to micro-resistance equations (Eqs. (7) and (12)) at cutter-edge radius in order to validate plowing resistance increasing than shearing, resulted in cutting temperature raising occurred at plowing zone. Simultaneously, the parameter setup of the rake angle at 5, 10, and 15° is presented to observe the relations of plowing zone, heat rate, von Mises stress, cutting force, and specific coefficients. The study did not consider the affection of the tool-HSS elastic deformation, and it needs to focus on the chip formation and shear deformation zone by the size effect. The size effect in micro-cutting process is relative to the theoretical model $R_{p,B} = R_{p,B}(R_{s,F}, R_{p,C}, \beta, \theta)$. The plowing effect at B point is greater than plowing at C and shearing at F if the micro-specific coefficient is close to 1. Hence, another important factor is differential-cutter angle γ relative to undeformed chip thickness if the micro-specific coefficient belongs to a reasonable range to avoid shearing-plowing coupled effect, such as the range $0 \leq \hat{K}_r \leq 0.9$, where $\hat{K}_r = 0$ is similar to the vertical feed on the workpiece or digging process. The variation of differential-cutter angle γ will affect the cutting force because the variation of the resultant force comes from the two component forces $md\dot{x}$ at x-direction and $md\dot{f}$ at y-direction. The resultant force vectors will affect the variation of undeformed chip thickness according to the theoretical model and demonstration in **Figures 4–6**. Therefore, the micro-cutting theory by the size effect is different from the traditional plastic theory. In the micro-cutting process, the tool and workpiece are usually symmetric in the z-direction. In brief, the model can be simplified from three-dimensional to two-dimensional in order to reduce the calculation as shown in **Table 1** and **Figure 8**. The influence of the factor affecting the variation of chip thickness comes from the variation of the resultant force at differential-cutter angle. If the radius of micro-cutter is larger, the resistance at x-direction will become larger due to the contacted areas for arc DE and arc BC increasing at the tool-

Material and tool geometry parameters			
Tool	HSS		
Workpiece	Al-7075		
Length*height (2D)	0.005 mm*0.002 mm		
Cutter-edge radius	0.00005 mm		
Rake angle (°)	5	10	15
Clearance angle (°)	10		
Process parameters			
Feed	0.0005 mm		
Room temp. (°C)	20		
Initial temp. of the workpiece (°C)	20		
Cutting depth (average chip thickness)	0.001 mm		
Cutting condition	Dry cutting		
Cutting velocity (mm/min)	30		
Friction coefficient	0.4–0.5		
Cutter condition	Single-cutter; orthogonal cutting		

Table 1.
Simulated parameters.

workpiece interface, and furthermore, the friction and resistance also increase, resulting in the differential-cutter angle varying with increasing resultant force, cutting force increasing with increasing resultant force if the cutting velocity was constant in the x-direction, and finally, the chip thickness varying with variation in the cutting force. From the observation of **Figures 5** and **6**, the resultant force indeed varied with the differential-cutter angle. Hence, the micro-specific coefficient $\hat{K}_r$ varied with the differential-cutter angle, and it was an important factor for the production precision in the machining process.

3.1 Micro-MDOF cutting dynamic simulation

The initial conditions for the material and workpiece as shown in **Table 1** and **Figure 8** can be set up by H-adaptive model fine mesh raising the cutting precision. Von Mises stress can be expressed as the wear on tool flank or the cutting resistance under plastic stress and friction. **Figure 9** demonstrated that the analysis of vectors on cutter-edge radius can prove that plowing at B (934 MPa) is larger than shearing at F (800 MPa) for (c) compared with (b) and (a); (a) theoretical model FBD for micro cutter-edge radius; (b) defined plowing and shearing on cutter radius, where the plowing angle is θ. The shearing and plowing resistance model can be established to predict the influence of tool geometry, rake angle, and plowing angle on the stress of the plowing zone and shearing zone. On the other hand, it can explain why cutting force increased, friction heat increased, and tool flank wear

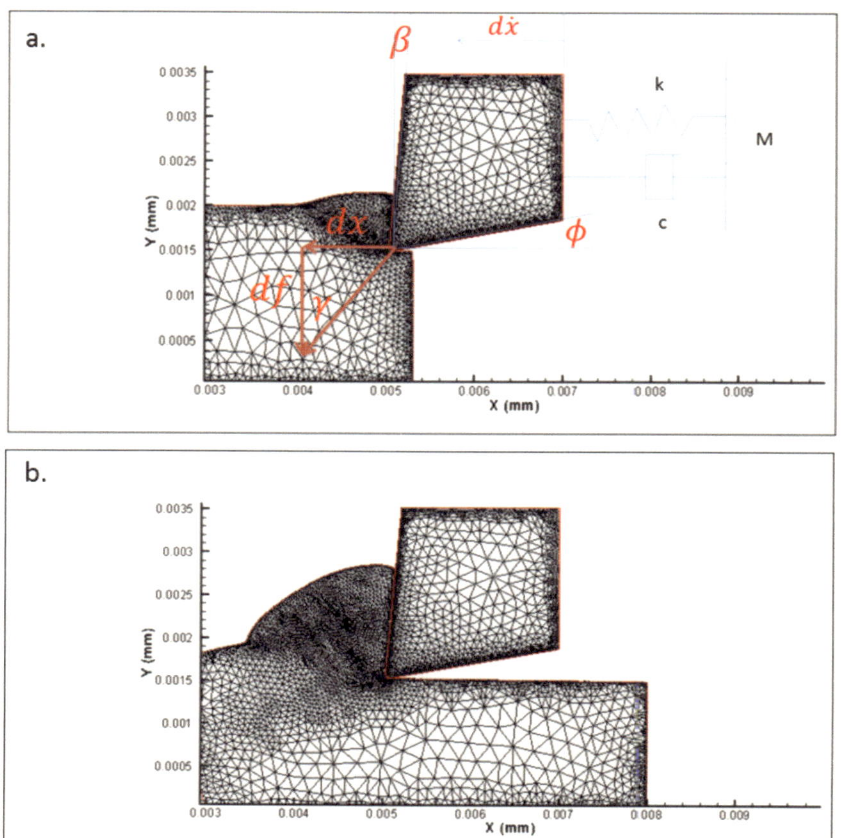

Figure 8.
H-adaptive meshing solid-model; (a) the cut-in process for micro-cutter and (b) the cutting process for steady-state chip formation.

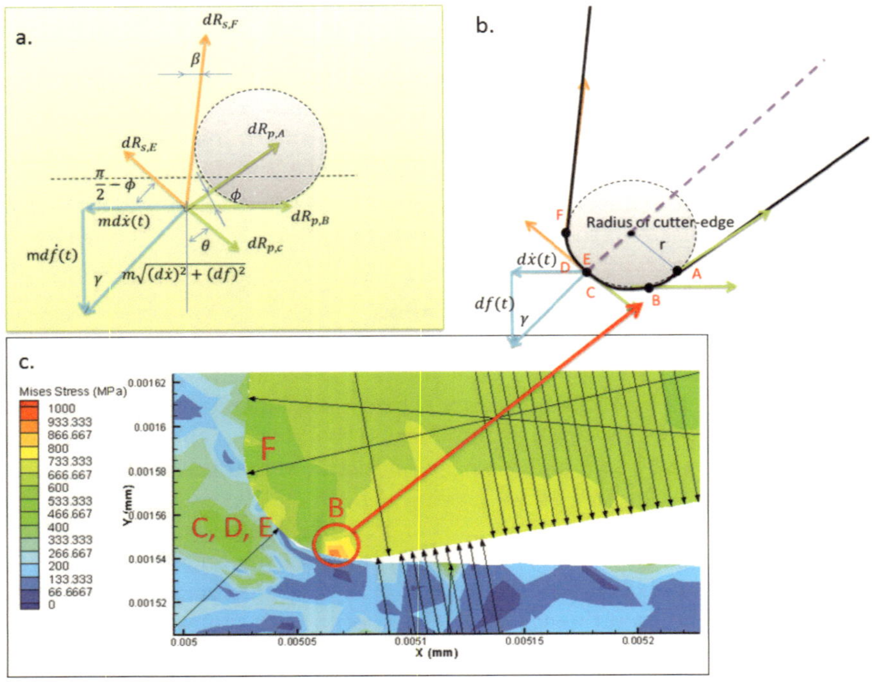

Figure 9.
The analysis of vectors on the cutter-edge radius can prove that plowing at B (934 MPa) is larger than shearing at F (800 MPa) for (c) compared with (b) and (a); (a) theoretical model FBD for micro-cutter-edge radius; (b) defined plowing and shearing on cutter radius; plowing angle is θ.

occurred by the size effect in the micro-cutting process. The finding results offer the importance for micro-MDOF cutting dynamics differed from the traditional cutting process. According to the simulated micro-cutting conditions of cutter-edge radius, the finding results are as follows: the maximum von Mises 934 MPa occurred on tool flank; the maximum heat rate $9.5E^6$ W/mm^2 occurred on the plowing zone to a part of the shearing zone; the maximum strain rate $9.5E^6$ occurred on the plowing zone to a part of the shearing zone similar to heat rate; and the higher temperature 43.4°C on the whole cutter-edge radius and shearing-plowing zone suffering size effect (**Figure 10**).

3.2 The relations of rake angle, cutting force, and cutting temperature

The results are nicely reasonable for workpiece Al-7075 and tool HSS. The variation of micro-cutting force is not uniform because of the roughness increasing on the workpiece surface after micro-cutting. The result can be proved by the plowing influence of size effect according to Eq. (7). The optimal process included a cut of depth of 0.001 mm, a cutting length of 0.003 mm, and a cutter edge of 38°C on workpiece Al-7075; the optimal cutting force in x-axis was 0.0005 N (Avg.) and the optimal cutting force in y-axis was 0.00028 N (Avg.) for better surface roughness Ra = 0.16. The higher temperature was 42.16°C on workpiece and tool HSS, and the maximum strain rate that occurred on the chip shearing zone was 9.33E06 (/s). The variation of rake angle compared with cutting force Fx, Fy, and cutting temperature presented that the rake angle 15° has better longer tool life because the cutting force is smaller than others as shown in **Figure 11**. The results can also be expressed by the resistance at B smaller than others. The cutter with rake angle 15° has lower cutting temperature through a long-time micro-cutting process as **Figure 12**. The trends of different rake angles for cutting force and cutting temperature are the same. The trends are nonlinear because of the size effect. The future work can investigate the influence of plowing angle.

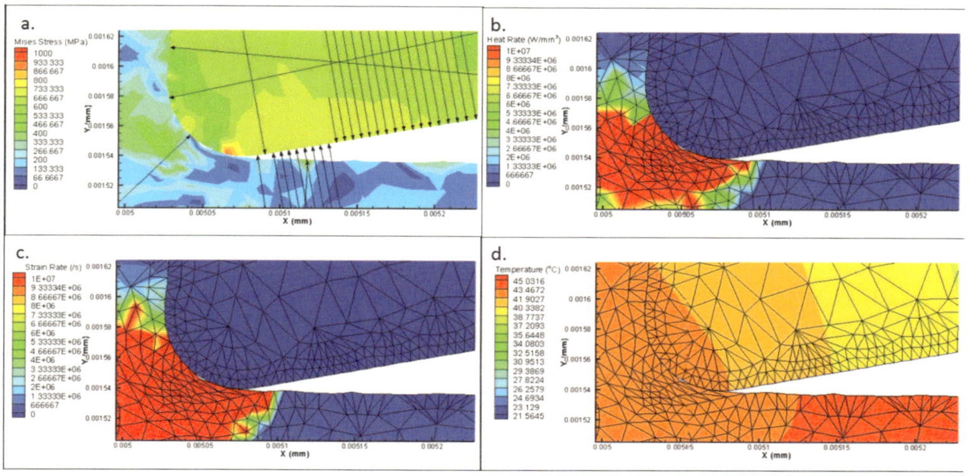

Figure 10.
The simulated micro-cutting conditions of cutter-edge radius. (a) The maximum von Mises 934 MPa occurred on tool flank; (b) the maximum heat rate 9.5E^{6} W/mm^{2} occurred on the plowing zone to a part of the shearing zone; (c) the maximum strain rate 9.5E^{6} occurred on the plowing zone to a part of the shearing zone similar to the heat rate; and (d) the maximum temperature 43.4°C on the whole cutter-edge radius and shearing-plowing zone suffering from size effect.

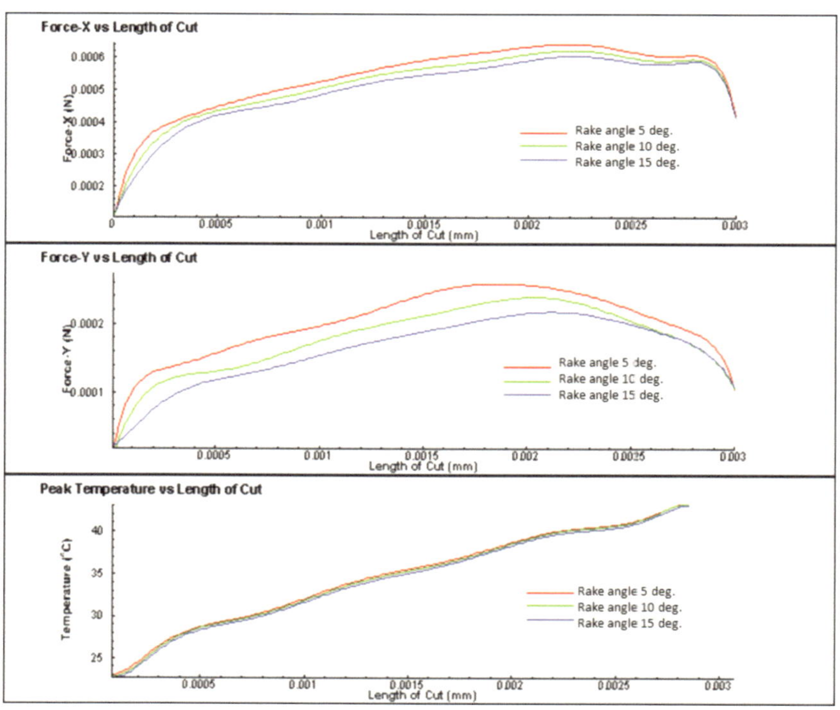

Figure 11.
The variation of rake angle compared with cutting force Fx and Fy and cutting temperature.

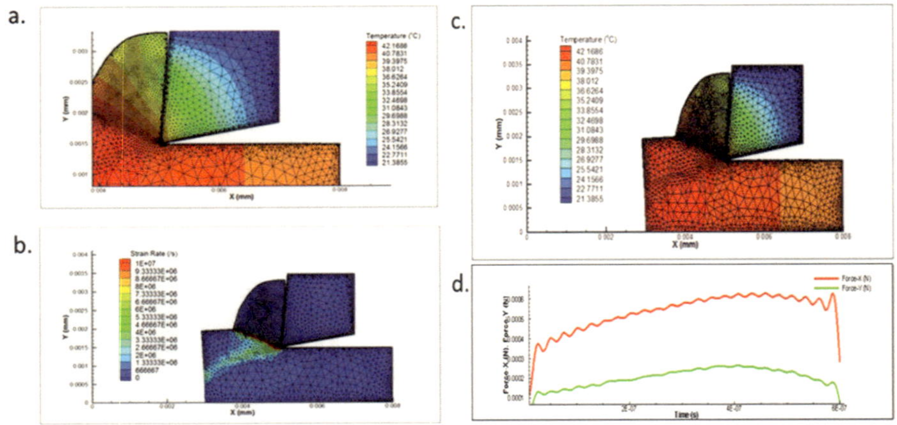

Figure 12.
Optimal process parameters in the study included a cut of depth of 0.001 mm, a cutting length of 0.003 mm, and a cutter edge of 38°C on workpiece Al-7075; the optimal cutting force in x-axis was 0.0005 N (Avg.) and the optimal cutting force in y-axis was 0.00028 N (Avg.) for better surface roughness Ra = 0.16. The higher temperature was 42.16°C on workpiece and tool HSS, and the maximum strain rate that occurred on the chip shearing plane $\dot{\varepsilon}$ was 9.33E06 (/s), which obeyed the generalized cutting criterion by numerical analysis.

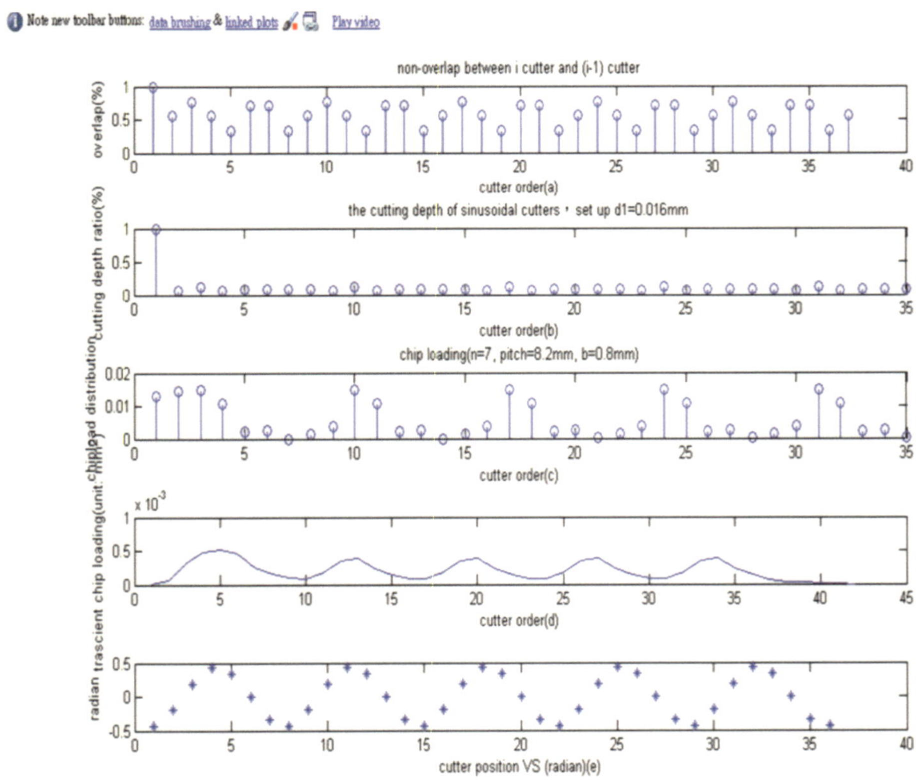

Figure 13.
Optimal geometric parameters for micro-cutting simulation for tool HSS and workpiece_Al-7075. (a) Non-overlap ratio for sinusoidal multi-cutters, (b) cutting depth ratio for sinusoidal multi-cutters, (c) discrete chip loading distribution, (d) continuous chip loading for multi-cutters, and (e) cutters' arrangement.

3.3 The simulation for multipoints and shearing and plowing

In order to express steady-state continuous chip formation, the cutting of multipoints has been simulated on the quasi-state cutting process as shown in

Figure 13 and the maximum chip loading of $0.56E^{-3}$ mm^2 as shown in **Figure 13(d)**. The non-overlap ratio and cutting depth ratio for cutters can be calculated by fractal geometry of the microscope. Through the assumed d1* = 0.125 mm of fractal mathematics, the results for multipoint cutting presented optimal geometric parameters for quasi-state cutting simulation of tool HSS and workpiece_Al-7075

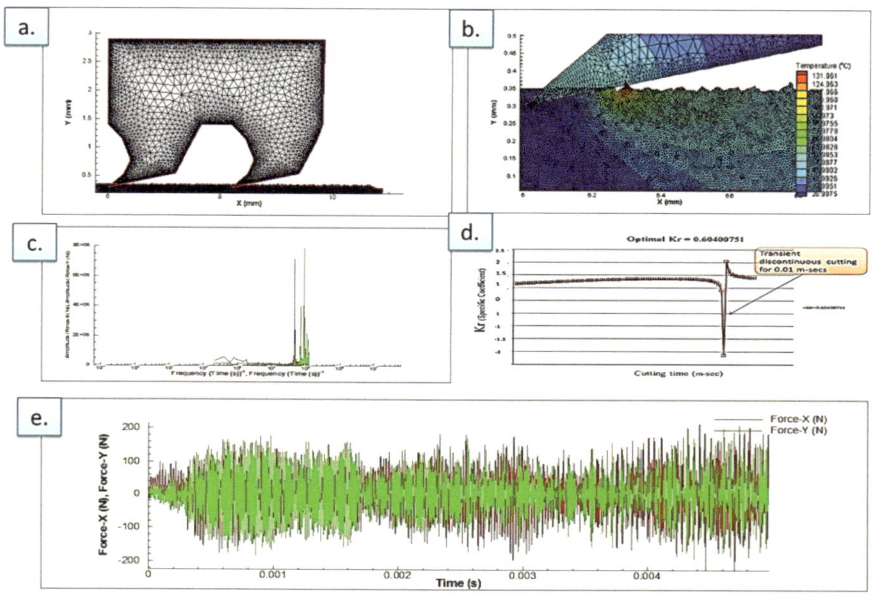

Figure 14.
Simulation for two cutters' cutting process [5]: (a) two cutters' modeling; (b) the effects on shearing equals plowing; (c) similar frequency for Fx and Fy; (d) specific coefficient Kr = 1.3 showed discontinuous characteristics; and (e) similar frequency for Fx *and* Fy *and cutting force distribution.*

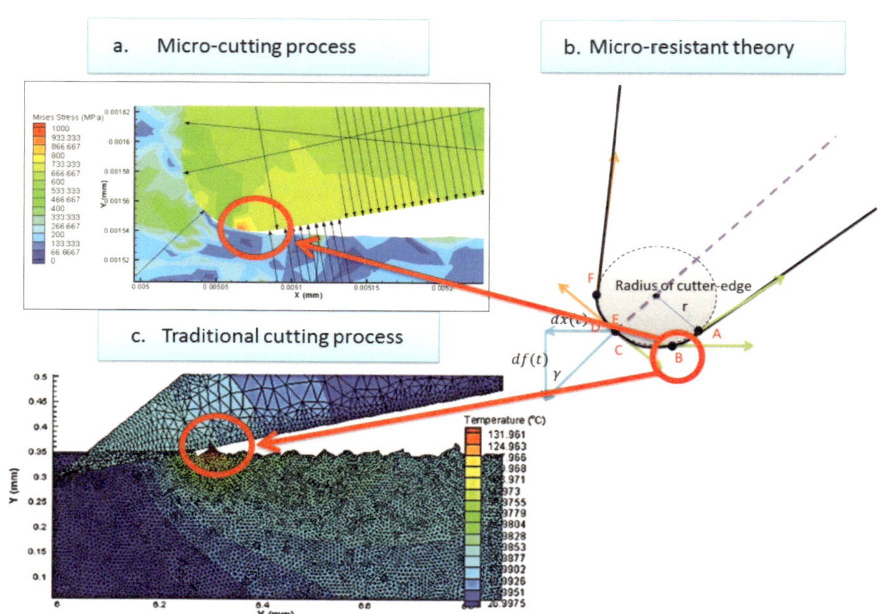

Figure 15.
To compare with micro-cutting, the traditional process has the same plowing effect, but the plowing in microscope effect is more obvious. (a) Cutting temperature in micro-cutting process, (b) micro-resistance theory, and (c) cutting temperature in the traditional process.

as shown in **Figure 13**: (a) non-overlap ratio for sinusoidal multi-cutters, (b) cutting depth ratio for sinusoidal multi-cutters, (c) discrete chip loading distribution, (d) continuous chip loading for multi-cutters, and (e) cutters arrangement. **Figure 13(d)** can be expressed as a continuous loading trend for unit period cutting similar to the trend of single cutter in the micromachining process without considering the influence of the size effect. Another simulation is to prove that the plowing and shearing coupling existed while $\frac{d\dot{x}(t)}{d\dot{f}(t)} = \hat{K}_r$ is close to 1 by traditional tool design in the cutting process. The special tool with two cutters of larger cutter-edge radius has simulated Kr to be close to 1 under the traditional cutting as shown in **Figure 14(a)** and **(b)**. The specific coefficient Kr = 1.3 showed discontinuous characteristics, where the traditional specific coefficient defined as $K_r = \overline{F}_x/\overline{F}_y$ and $\overline{F}_x, \overline{F}_y$ means the average cutting force. **Figure 14(e)** shows similar frequency for F_x and F_y and cutting force distribution. The excited-vibration of the tool will occur when the micro-specific coefficient $\hat{K}_r$ is close to 1. To compare with micro-cutting, the traditional cutting process has the same plowing effect, but the plowing in microscope effect is more obvious as shown in **Figure 15**. The validation can be proven as the derived theory agreed well with the simulation in the micro-cutting process.

4. Conclusions

The major results have been summarized as follows:

1. The optimal process included a cut of depth of 0.001 mm, a cutting length of 0.003 mm, and a cutter edge of 38°C on workpiece Al-7075; the optimal cutting force in x-axis was 0.0005 N (Avg.) and the optimal cutting force in y-axis was 0.00028 N (Avg.) for better surface roughness Ra = 0.16. The higher temperature was 42.16°C on workpiece and tool HSS, and the maximum strain rate that occurred on the chip shearing zone was 9.33E06 (/s), which obeyed the generalized cutting criterion by numerical analysis.

2. For the steady-state chip formation in the micro-cutting process, the rake angle is usually constant, but the plowing angle is relative to the tool-edge precision and surface roughness on the workpiece.

3. While micro-specific coefficient $\hat{K}_r$ is close to 1, the plowing zone will increase friction, stress, resistance, and even cutting excited-vibration or chipping, resulting in discontinuous chipping.

4. The variation of the rake angle will affect the cutting force and plowing zone according to Eq. (7). The average micro-cutting force F_x = 0.5 mN and F_y = 0.22 mN and micro-specific coefficient $\hat{K}_r$ = 0.44; the maximum cutting temperature is distributed at 35–40°C.

5. Through fractal mathematics, the results presented optimal geometric parameters for micro-cutting simulation for tool HSS and workpiece_Al-7075: overlap ratio between cutters, average cut-depth ratio between cutters, and chip load (undeformed chip formation areas and shapes) distribution on cutter order.

6. The study proposed the mathematical model of micro-cutting resistance to predict the conditions at cutter-edge radius.

7. The average micro-cutting force Fx = 0.5 mN and Fy = 0.22 mN and specific coefficient Kr = 0.44; the maximum cutting temperature is distributed at 35–40°C. To compare the variation of rake angle, rake angle at 15° has smaller micro-cutting force, and hence, the tool design has longer tool life in the micro-cutting process.

8. The study developed the micro-MDOF cutting dynamics model and micro-fractal equation to simulate the micro-cutting process.

9. Due to suffering from the size effect of cutter-edge radius r in the micromachining process, the differential-cutter angle is relative to two factors: plowing angle and rake angle. For the steady-state chip formation in the micro-cutting process, the rake angle is usually constant, but the plowing angle is relative to the tool-edge precision and surface roughness on the workpiece.

10. From the view of the micro-cutting process, the definition of micro-specific coefficient is $\hat{\mathrm{K}}_{\mathrm{r}} = \frac{d\dot{x}}{d\dot{f}}$, and the definition of average micro-specific coefficient is $\hat{\mathrm{K}}_{\mathrm{r,\,avg}} = \frac{d\dot{x}_2 - d\dot{x}_1}{d\dot{f}_2 - d\dot{f}_1} = slope$.

11. To compare with micro-cutting, the traditional cutting process has the same plowing effect, but the plowing in microscope effect is more obvious as shown in **Figure 15**. The validation can be proven as the derived theory agreed with the simulation in the micro-cutting process.

Acknowledgements

This work was supported in part by Taiwan NSC under Grant No. MOST 105-2221-E-327-015 and the industrial plan—Development of Ultra Speed Intelligent CNC Band Saw Machine from No. 105RB07. Special thanks to Prof. Ching-Hua Wei, Prof. Chin-Tu Lu, Prof. Jung-Zen Huang, One-on-One group members and my good friend Sam Fang for supporting my study; Prof. Sheng-Jye Hwang, Prof. Ta-Hui Lin, and Prof. Rong-Shean Lee at NCKU for their support in the process of this study; and Prof. Yunn-Lin Hwang and Prof. Jeng-Haur Horng for their research cooperation at the National Formosa University.

Author details

Sung-Hua Wu
Department of Mechanical Engineering, National Cheng Kung University, Tainan City, Taiwan (R.O.C)

*Address all correspondence to: jackywoo0923@gmail.com

References

[1] Mandelbort BB. The Fractal Geometry of Nature. New York: W.H. Freeman; 1982

[2] Cao ZY, He N, Li L. A finite element analysis of micro/meso-scale machining considering the cutting edge radius. Applied Mechanics and Materials. 2008; **10–12**:631-636. DOI: 10.4028/www.scientific.net/AMM.10-12.631

[3] Wu S-H, Junz Wang J-J, Lee R-S. Chip fractal geometry and loading characteristics of sinusoidal multi-cutters in hack-sawing process. International Journal of Machine Tools & Manufacture. 2012;**59**:65-80. DOI: 10.1016/j.ijmachtools.2012.01.005

[4] Wu S-H, Huang M-S, Jhou C-E, Wei C-C. Study on the cutting efficiency of high-speed band saw blade by Taylor tool life and fractal equations. In: MATEC Web of Conferences. Vol. 201. 2018. p. 01001. DOI: 10.105/matecconf/201820101001

[5] Sung-Hua W. Optimal process parameters of shearing and plowing by fractal equation in micro-cutting strategy. In: 2016 International Conference on System Science and Engineering (ICSSE); Puli, Taiwan. DOI: 10.1109/ICSSE.2016.7551643

[6] Adamsa DP, Vasileb MJ, Benavidesa G, Campbella AN. Micromilling of metal alloys with focused ion beam–fabricated tools. Journal of the International Societies for Precision Engineering and Nanotechnology. 2001;**25**:107-113

[7] Dib MHM, Duduch JG, Jasinevicius RG. Minimum chip thickness determination by means of cutting force signal in micro end milling. Precision Engineering. January 2018;**51**:244-262. DOI: 10.1016/j.precisioneng.2017.08.016

[8] Chae J, Park SS, Freiheit T. Investigation of micro-cutting operations. International Journal of Machine Tools & Manufacture. 2006; **46**:313-332. DOI: 10.1016/j.ijmachtools.2005.05.015

[9] de Oliveira FB, Rodrigues AR, Coelho RT, de Souza AF. Size effect and minimum chip thickness in micromilling. International Journal of Machine Tools & Manufacture. 2015;**89**: 39-54. DOI: 10.1016/j.ijmachtools.2014.11.001

[10] Zhang X, Ehmann KF, Yu T, Wang W. Cutting forces in micro-end-milling processes. International Journal of Machine Tools & Manufacture. 2016; **107**:21-40. DOI: 10.1016/j.ijmachtools.2016.04.012

[11] Tansel IN, Arkan TT, Bao WY, Mahendrakar N, Shisler B, Smith D, et al. Tool wear estimation in micro-machining. Part I: Tool usage–cutting force relationship. International Journal of Machine Tools & Manufacture. 2000; **40**:599-608

[12] Tansel IN, Arkan TT, Bao WY, Mahendrakar N, Shisler B, Smith D, et al. Tool wear estimation in micro-machining. Part II: Neural-network-based periodic inspector for nonmetals. International Journal of Machine Tools & Manufacture. 2000;**40**:609-620

[13] Malekian M, Mostofa MG, Park SS, Jun MBG. Modeling of minimum uncut chip thickness in micro machining of aluminum. Journal of Materials Processing Technology. 2012;**212**: 553-559. DOI: 10.1016/j.jmatprotec.2011.05.022

[14] Wu S-H, Lee R-S, Junz Wang J-J. Specific energy distributions for sinusoidal multi-cutters in groove-sawing process. Journal of Materials Processing Technology. 2013;**213**: 641-659. DOI: 10.1016/j.jmatprotec.2012.11.019

[15] Afazov SM, Zdebski D, Ratchev SM, Segal J, Liu S. Effects of micro-milling conditions on the cutting forces and process stability. Journal of Materials Processing Technology. 2013;**213**: 671-684. DOI: 10.1016/j.jmatprotec.2012.12.001

[16] Filiz S, Ozdoganlar B. A three-dimensional model for the dynamics of microendmills including bending, torsional and axial vibrations. Precision Engineering. 2011;**35**:24-37

[17] Filiz S, Conley C, Wasserman M, Ozdoganlar B. An experimental investigation of micro-machinability of copper 101 using tungsten carbide micro-endmills. International Journal of Machine Tools and Manufacture. 2007; **47**:1088-1100

[18] Afazov SM, Ratchev SM, Segal J, Popov AA. Chatter modelling in micro-milling by considering process nonlinearities. International Journal of Machine Tools & Manufacture. 2012;**56**: 28-38. DOI: 10.1016/j.ijmachtools.2011.12.010

[19] Afazov SM, Ratchev SM, Segal J. Prediction and experimental validation of micro-milling cutting forces of AISI H13 stainless steel at hardness between 35 and 60 HRC. International Journal of Advanced Manufacturing Technology. October 2012;**62**(9-12):887-899. DOI: 10.1007/s00170-011-3864-7

[20] Guo Q, Zhao B, Jiang Y, Zhao W. Cutting force modeling for non-uniform helix tools based on compensated chip thickness in five-axis flank milling process. Precision Engineering. 2018;**51**: 659-681. DOI: 10.1016/j.precisioneng.2017.11.009

[21] Malekian M, Park SS, Jun MBG. Modeling of dynamic micro-milling cutting forces. International Journal of Machine Tools & Manufacture. 2009; **49**:586-598. DOI: 10.1016/j.ijmachtools.2009.02.006

Chapter 7

Multiscale Simulation of Surface Defect Influence in Nanoindentation by a Quasi-Continuum Method

Zhongli Zhang, Yushan Ni, Jinming Zhang, Can Wang and Xuedi Ren

Abstract

Microscopic properties of crystal aluminum thin film have been investigated using the quasi-continuum method in order to study the influence of surface defects in nanoindentation. Various distances between the surface pit defect and indenter and various sizes of the pit have been calculated. In this simulation, as the distance between the pit and indenter increases, the nanohardness increases in a wave that goes up in a period of three atoms, and it is found closely related to the crystal structure of periodic atom arrangement on {1 1 1} atomic close-packed planes of FCC metal; there is almost no influence on the nanohardness when the adjacent distance between the pit and indenter is more than 16 atomic spacing. We have modified the theoretical equation of the necessary load for elastic-to-plastic transition of Al film with the initial surface defect size. Furthermore, when the size coefficient of width (of height) equals about one unit (half unit), the yield load experiences an obvious drop. When it reaches about two units (one unit), the yield load is nearly close to that of the nanoindentation on a stepped surface. Additionally, compared to the width, the height of surface pit defect displays a greater influence on the yield load of thin film.

Keywords: surface pit defect, multiscale simulation, size effect, distance effect, quasi-continuum method

1. Introduction

Nanoindentation [1], relatively simple and effective, has already been used as a standard technique for evaluating mechanical properties of thin films [2–7]. In the recent years, a number of scientists have focused on thin films with defects through simulations and experiment [8–11]. Yu has analyzed the effects of nanocavity on nanoindentation, which is one kind of defect [12]. Surface roughness, grouped by the pits and steps, has already been a popular topic, and lots of nanoindentation simulations on a step have been studied [13–20]. It is well-known that the surface pits are very common in polycrystalline surfaces on microchips or microelectromechanical systems (MEMS). Therefore, it is necessary and significant to observe the nanoindentation on the pitted surface. Ni et al. [21] have compared nanoindentation of Al thin film with and without surface defects by multiscale simulation, and Zhang et al. [22] have probed the delay effect of dislocation nucleation in nanoindentation due to the surface pit defect. However, the distance effect of the pit and the size effect on elastic-plastic transition, which is especially important to thin film performance in microchips and MEMS, have not been taken into account yet. Now, this chapter is further to study and represent the distance effect [23] and the size effect of the pit [24] on nanohardness by quasi-continuum (QC) method [25], which is a relatively effective way to investigate large-scale model, where the molecular dynamics (MD) is mostly limited by the model scale.

2. Distance effect of surface pit defect

2.1 Methodology

The quasi-continuum (QC) method [26] is an effective mixed continuum and atomistic approach for simulating the mechanical response, especially in large-scale materials. The Ercolessi-Adams potential, which is one of the EAM potentials [27–30], is used in this QC method, in order to describe the atomistic behavior. The parameters for Al in this potential are as shown in **Table 1**.

The schematic illustration of nanoindentation model with a pit defect and its unit cell model of Al in the selected directions is as shown in **Figure 1**, where the width of the rigid rectangular indenter is 0.932 nm (four times the lattice constant of Al in [1 1 1] direction (d_0)). Such indenter width is chosen based on the simulation example "Nano-Indentation by a Square Punch" in QC tutorial document [32] and others' works [21, 22]. It is convenient to simulate and analyze that if the indenter shape is set rectangular, result from that the boundary of energy field (displacement field) and the distance between the pit and the indenter will not changed during the indenter being driven down into the $(\bar{1}10)$ surface in this simulation, and that is exactly necessary to probe the distance effect of the pit. The width D and depth H of the pit in the model are 0.688 and 0.730 nm, respectively. Such pit size is chosen relatively moderate and proper, more sensitive to the distance effect

Content	Value
Crystallographic lattice (a_1)	0.4032 nm
Atomic lattice spacing in [1 1 1] direction (d_0)	0.2328 nm
Atomic spacing in $[\bar{1}1\ 0]$ direction (h_0)	0.1426 nm
Burgers vector ($\vec{b}$)	0.285 nm
Shear modulus (μ)	33.14 GPa
Poisson (ν)	0.319
(1 1 1) surface energy (γ_{111})	0.869 J/m^2 [a]
C_{11}	117.74 GPa [b]
C_{12}	62.06 GPa [b]
C_{44}	36.67 GPa [b]

[a](1 1 1) surface energy γ_{111} is 0.869 J/m^2, which is comparable with the experimental values of 1.14–1.20 J/m^2 [31].
[b]The experimental values extrapolated to T = 0 K are C_{11} = 118.0 GPa, C_{12} = 62.4 GPa, and C_{44} = 32.5 GPa [27].

Table 1.
The parameters for Al in EAM potential.

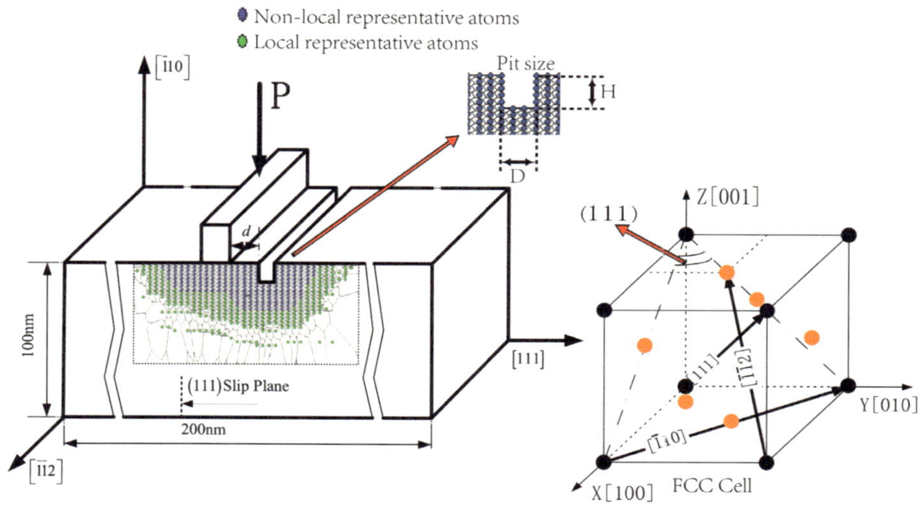

Figure 1.
The schematic illustration of nanoindentation model with a pit defect and its unit cell model of Al in the selected directions, where the unusual shapes in local region are not finite elements, they are just the schematic of its specific region that one corresponding representative atom belongs to.

based on our previous works [21, 22] and some pre-simulations. The thickness of this model in the out-of-plane direction is 0.4938 nm, which equals to the minimal repeat size applying the periodic boundary condition. Fifteen different d (defined as the adjacent distance between the pit and indenter) have been calculated in this chapter, respectively, $1d_0$, $2d_0$, $3d_0$, $4d_0$, $5d_0$, $6d_0$, $7d_0$, $8d_0$, $9d_0$, $10d_0$, $11d_0$, 12d0, $13d_0$, $17d_0$, and $21d_0$, in order to make a more comprehensive investigation.

The parameter "*PROXFACT*" in QC method is applied to judge the range of nonlocal effects. The repatom is made nonlocal when any element is in its range of *PROXFACT*rcut*, where *rcut* is defined as the atomistic potential cutoff radius. Nonlocality is triggered if $\max_{a,\,b;\,k}\left|\lambda_k^a - \lambda_k^b\right| > epscr$ (k = 1, 2, 3), where λ_k^a and λ_k^b are the eigenvalues of right Cauchy-Green deformation tensor $\mathbf{C}$ in two elements a and b, and the factor "*epscr*" is applied to judge whether a repatom has to be made nonlocal because of the significant variations in the deformation gradients around the repatom.

For normal atomistic modeling standards, the dimensions of this simulation thin film is quite large with approximately 1.3 million atoms or 4 million degrees of freedom (as shown 0.1 μm in height and 0.2 μm in width in **Figure 1**). Fortunately, QC method applies the molecular dynamics model at the intense deformation region and a finite element model elsewhere in order to reduce the degrees of freedom without losing atomistic details, where only 4000 atoms or 12,000 degrees of freedom have to be treated in this model by comparison, and can be easily finished in a few days through personal computer.

2.2 Results

2.2.1 Nanoindentation without defect

Nanoindentation without defect is necessary to be studied for comparison, and the load-displacement curve of the nanoindentation on a defect-free surface is shown in **Figure 2**, where load (N/m) is presented as per unit length of indenter in the out-of-plane direction. It can be easily found out that the load curve gradually

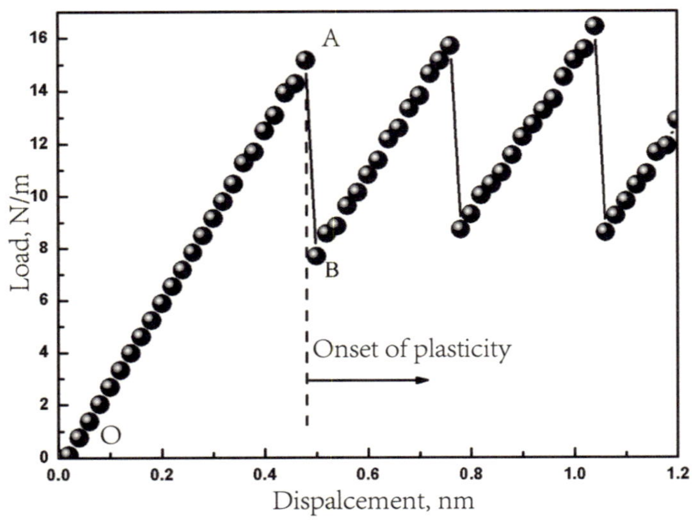

Figure 2.
Load-displacement curves for nanoindentation on Al film without surface pit.

increases in the loading process (OA segment), indicating the elastic stage of thin films. The load first reaches the maximum value of 15.14 N/m at the load step 0.48 nm (point A) and then suddenly drops to the minimum value of 7.67 N/m at point B.

To find out the reason of such abrupt load decline in AB segment, the atoms structure and out-of-plane displacements have been probed and shown in **Figure 3**, where the step of 0.48 and 0.50 nm is, respectively, corresponding to the point A and point B in **Figure 2**.

Through **Figure 3**, we can make a conclusion here that the load reaches the critical value for dislocation emission at point A, which indicates the onset of the plastic stage. After that two Shockley partial dislocations are emitted at point B. Therefore the nanohardness of Al thin film without defect is 16.24 GPa, calculated by equation [33]: $H = \frac{P_{max}}{A}$, where P_{max} is the maximum value of load and A is the contact area of the indenter.

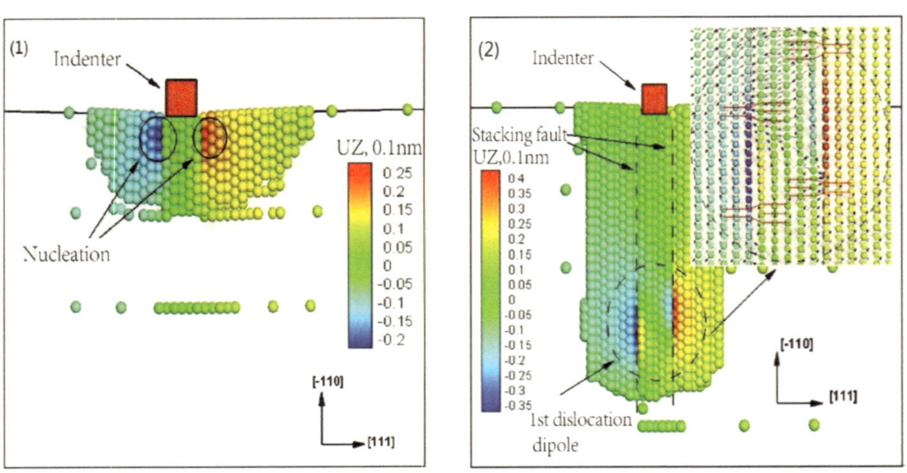

Figure 3.
*Snapshot of atoms under indenter and corresponding out-of-plane displacement plot, where UZ is atom displacement at out-of-plane. (1) point A in **Figure 2** (dislocation nucleation); (2) point B in **Figure 2** (dislocation emission).*

Due to the indenter width 0.932 nm in this simulation, the yield load is obtained approximately 15.14 N/m, which is smaller than 24.7 N/m acquired by Tadmor and Miller [31] with the indenter width 2.5 nm. It is reasonable that the paper [34] shows that as the indenter width decreases, the yield load would decline because of the requirement decline of the necessary strain energy.

2.2.2 Distance effect of the surface pit

Figure 4 shows the nanohardness with various distances cases of the surface pit defects. Compared to the defect-free situation shown as a red horizontal, it indicates that the nanohardness of pitted surface has been declined. That is because a discontinuity at the boundary and the structure may lead to the reduction of stain energy storage when indenting. Additionally, when the adjacent distance between the pit and indenter (*d*) increases, the nanohardness increases in a wave that goes up in a period of three atoms (donated by circle in **Figure 4**) and finally tends to the case of nanoindentation on a defect-free surface.

To make a further probe, it is well-known that many physical properties mostly depend on the stacking patterns of atoms, such as cleavage, electronic band structure, and optical transparency [35]. Based on this simulation, the periodic arrangement of atoms "ABCABC" on {1 1 1} atomic close-packed planes of face-centered cubic metal (the illustration as shown in **Figure 4**) is exactly corresponding to increasing distances *d* on [1 1 1] direction. That is to say, when the pit moves each atom in the [1 1 1] direction away from the indenter, the strain energy at the pit surface on the {1 1 1} stacking fault energies (SFE) changes because of the proximity of the pit [36]. Consequently, such wave pattern associated with a cycle of three atoms is closely related to the crystal structure of periodic atom arrangement on {1 1 1} atomic close-packed planes of FCC metal.

A further discussion has been made in order to figure out the spatial extent of surface pit influence on nanohardness. It can be found out from **Figure 4** that when the distance between the pit and indenter increases, the nanohardness gradually close to the nanohardness of defect-free case (16.24 GPa). If we set 1.5% to determine whether the nanohardness influence exists, it can be found out that when the adjacent distance (*d*) goes 16 atomic spacing far away from the indenter, there is almost no effect on nanohardness (as shown in **Figure 4**). Moreover, it can be

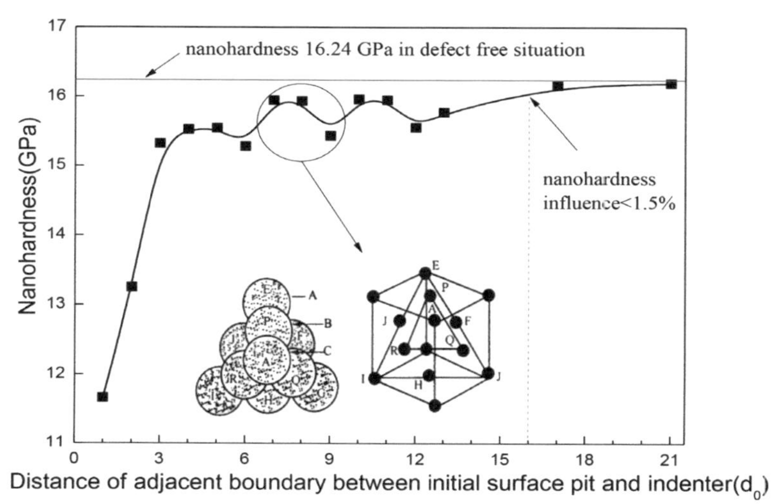

Figure 4.
Nanohardness vs. distance of adjacent boundary between the pit and indenter.

predicted that each material has its critical value of such spatial extent of surface pit influence on nanohardness, which might have great significance to the size design of thin film in microchips without obvious reduction of the hardness.

However, it can be easily found out that the first three distances cases, respectively, $d = 1d_0$, $d = 2d_0$, and $d = 3d_0$, do not match such wave pattern well. Atomic structure and corresponding strain distribution of these three cases have been further carried out to explain such unusual phenomenon.

Von Mises strain distribution and a strain comparison before and after the notch propagation in the distance cases of, respectively, $1d_0$, $2d_0$, and $3d_0$, are shown in **Figure 5**. It can be easily found out that a notch formed at the left side of surface pit in the distance cases of $1d_0$ and $2d_0$, which actually induces serious damage to the structure of materials and severe strain concentration (as shown in **Figure 5A–D**), while it does not if the distance d equals $3d_0$ (as shown in **Figure 5E** and **F**). That is to say, due to the great reduction of the nanohardness in the cases of $1d_0$ and $2d_0$, the first three distance cases in nanohardness curve as shown in **Figure 4** will not match the wave pattern that goes up in a period of three atoms.

2.2.3 Critical load for dislocation emission with initial surface pit

As is known to all that the pit influences the nanohardness is actually through the way of affecting nucleation and emission of the dislocation. It is necessary and significant to make a further probe on the critical load for elastic-to-plastic transition in the case of nanoindentation on the pitted surface according to the formula of defect-free model, where the formula to calculate the critical load for dislocation emission is carried out by Tadmor [31], shown as Eq. (1):

$$P_{cr} = \frac{\mu b}{4\pi(1-v)} \ln \frac{32h(h+2a)a^2}{b^4} + 2\gamma_{111} + \frac{1}{2}kb \tag{1}$$

where P_{cr} is the critical load of dislocation emission, h is the depth of dislocation dipole emitted down beneath the indenter, a is the half width of indenter, γ_{111} is the (1 1 1) surface energy of Al single crystal, and k is the slope of the load-displacement curve during the elastic section.

The critical load calculated by the simulation results of QC method and by Eq. (1) of dislocation theory has been displayed, respectively, as "QC data" and "theory load" in **Table 2**. It is necessary to note that the simulation data in the case of $1d_0$ and $2d_0$ is not suitable to be taken into account because of the notch propagation. The differential of the critical data between QC method and dislocation theory is fluctuant as the distance (d) changes. Based on the result that the nanohardness goes up in a period of three atoms, the critical load for elastic-to-plastic is also in such periodicity. Thus, the correction form (set as Δ) might be reasonably defined as the following:

$$\Delta = A(d) + B \cdot Sin\,(d) \tag{2}$$

where $A(d)$ is the hardness reduction due to the surface pit and $B{\cdot}Sin(d)$ is specially set for the periodic arrangement of atoms. It is well recognized that the critical load of elastic-to-plastic transition will decrease [37], when the pit size (D, H as shown in **Figure 1**) increases. We use a dimensionless factor $\frac{D}{a_1} \cdot \frac{H}{a_1}$ (dividing by the crystallographic lattice constant) to express the size influence of the pit, which has already been demonstrated reasonable in published article [24]. If the distance between

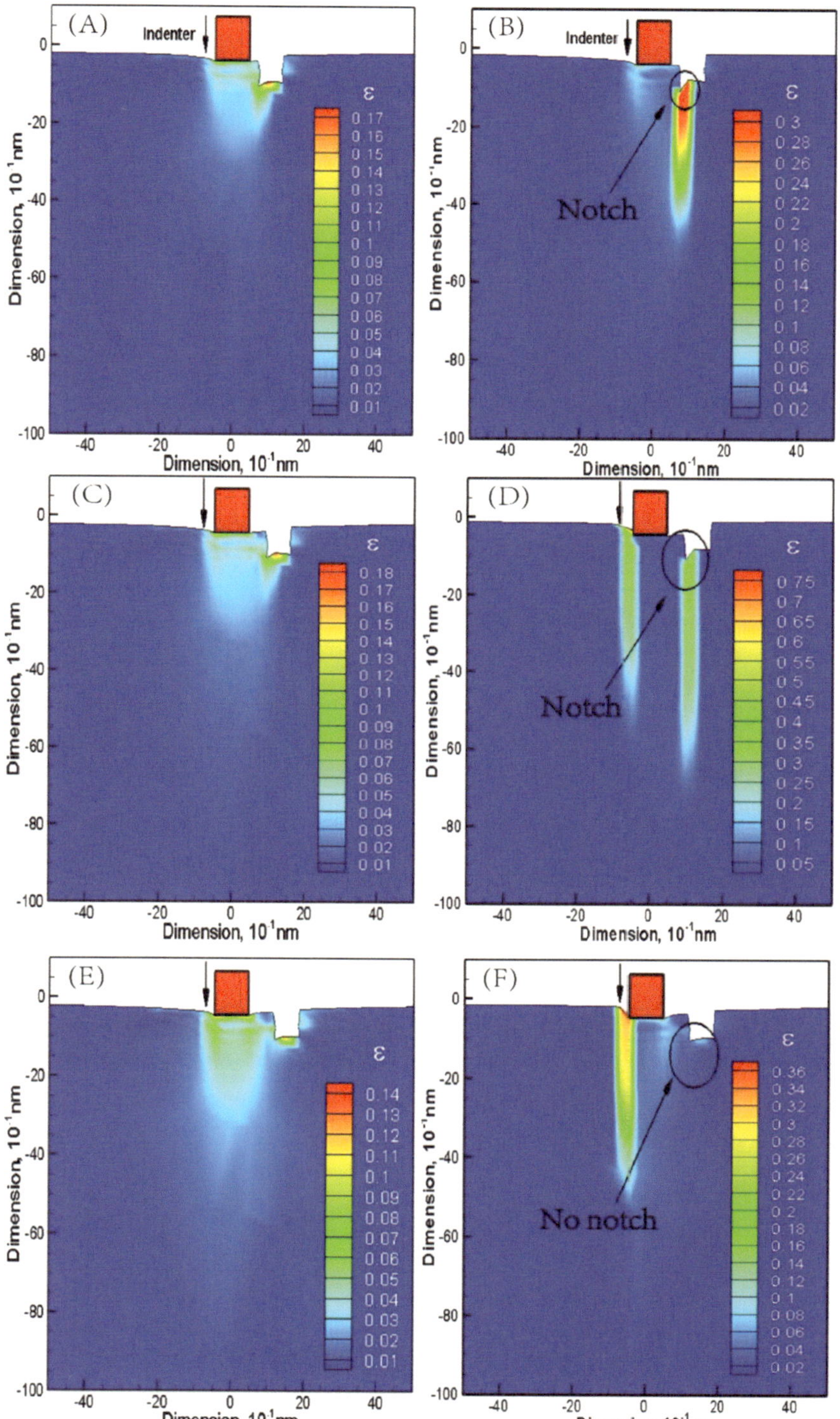

Figure 5.
Von Mises strain distribution of notch propagation. (A) $d = 1d_o$ at the load step of the indenter 0.38 nm; (B) $d = 1d_o$ at the load step of the indenter 0.4 nm; (C) $d = 2d_o$ at the load step of the indenter 0.44 nm; (D) $d = 2d_o$ at the load step of the indenter 0.46 nm; (E) $d = 3d_o$ at the load step of the indenter 0.46 nm; (F) $d = 3d_o$ at the load step of the indenter 0.48 nm.

Distance (d_0)	QC data (N/m)	Theory load (N/m)	Data difference (N/m)
3	14.28	18.02	3.75
4	14.46	17.29	2.83
5	14.48	17.88	3.39
6	14.24	17.41	3.15
7	14.86	17.96	3.14
8	14.85	17.65	2.83
9	14.38	17.92	3.07
10	14.87	17.87	3.49
11	14.86	18.03	3.16
12	14.49	17.56	2.70
13	14.70	17.99	3.50
17	15.06	18.04	3.34
21	15.09	18.17	3.11

Table 2.
The comparison of critical load between QC method and dislocation theory.

the indenter and the pit is infinitely large, the influence on nanohardness can be almost ignored, and if the pit size increases, the rate of nanohardness change decreases with the distance variation. According to the function property and calculation formula designed by Tadmor [31], we take the form of $\ln\left(1+\left(\frac{d_0}{d}\right)^{\frac{d_0}{D}\cdot\frac{h_0}{H}}\right)$ to express the distance effect of surface pit, considering that it is relevantly reasonable. Moreover, the affection of surface pit is closely due to the material property such as Burgers vector $\vec{b}$, shear modulus μ, and Poisson ν. It is reasonable to apply $\frac{\mu b}{4\pi(1-\nu)}$ to express the influence of material property based on Eq. (1). In addition, the atomic periodical arrangement is actually three atoms "ABCABC" on {1 1 1} atomic close-packed planes of FCC metal. Namely, a form of $Sin\left(\frac{2\pi}{3d_0}\cdot d+\varphi\right)$ is proper to express the periodicity of atom arrangement. Besides, we know that the unit of correction term (Δ) is exactly N/m. So, the correction can be defined as the following based on the discussion above:

$$\Delta=\alpha\cdot\frac{\mu bDH}{4\pi(1-\nu){a_1}^2}\ln\left(1+\left(\frac{d_0}{d}\right)^{\frac{d_0}{D}\cdot\frac{h_0}{H}}\right)+\beta\cdot\frac{\mu b}{4\pi(1-\nu)}\cdot Sin\left(\frac{2\pi}{3d_0}\cdot d+\varphi\right)\quad(3)$$

where α, β, and φ are three constants that need to be matched and fitted. Based on the simulation data in **Table 2**, these three constants α, β, and φ can be acquired by calculation approximately $\frac{3}{2}$, $\frac{2}{15}$, and $-\frac{\pi}{3}$, respectively. So the critical load for the first dislocation emission of Al film has been revised with initial surface pit as follows:

$$P_{cr}{}^{*}=\frac{\mu b}{4\pi(1-v)}\ln\frac{32h(h+2a)a^2}{b^4}+2\gamma_{111}+\frac{1}{2}kb-\frac{3\mu bDH}{8\pi(1-\nu){a_1}^2}\ln\left(1+\left(\frac{d_0}{d}\right)^{\frac{d_0}{D}\cdot\frac{h_0}{H}}\right)-\frac{\mu b}{30\pi(1-\nu)}\cdot Sin\left(\frac{2\pi}{3d_0}\cdot d-\frac{\pi}{3}\right)\quad(4)$$

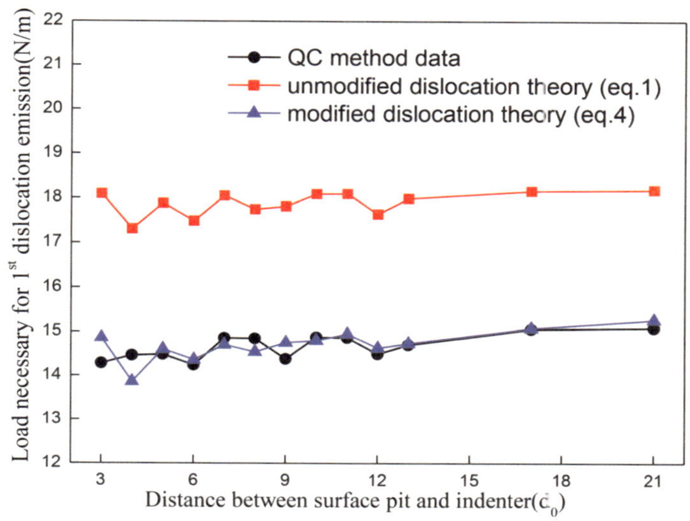

Figure 6.
The comparison of the necessary load for elastic-to-plastic transition of Al thin film with various distances between the pit and the indenter calculated by the theoretical formula before and after modification.

Figure 6 shows the comparison of the critical load for dislocation emission of Al thin film in different distance cases calculated by the theoretical formula before and after modification. Though there is no parameter d in Eq. (1), the curves with blocks are calculated by depth h corresponding each distance case in this simulation. The simulation QC data is closer to the theoretical results which are calculated by Eq. (4) after modification. That is to say, such modification to the theoretical formula is justified as the pit size and the distance between the pit and indenter have both been taken into account.

The modified formula displays the decreasing trend of nanohardness as the distance between the pit and indenter increases, which quite agrees with the experimental results of nanoindentation on the surface step with different distances [14]. Further, such study might be referential to the research of material properties with defects, especially in microchips and MEMS.

3. Size effect of surface pit defect

3.1 Method and model

The quasi-continuum method is adopted in this simulation, which is one of the multiscale approaches that keeps an atomistic description at highly deformed regions, whereas a linear elastic continuum method is implemented far away from this dislocation core. In this simulation, the Ercolessi-Adams potential (EAM) [38] is also applied to describe the atomistic behavior of the system.

The nanoindentation model used in the simulation has been shown in **Figure 7**, where the x-axis direction is [1 1 1] direction, the y-axis direction is $[\bar{1}\ 1\ 0]$, and the outer-of-plane z direction is $[\bar{1}\ \bar{1}\ 2]$ direction. And its schematic of local and nonlocal representative atoms with initial surface pit defect has been shown in **Figure 8**. Such orientation is selected to facilitate dislocation emission. The model size is 200 nm in width and 100 nm in height, which is about 10 times of the usual MD size. The width of the rigid indenter is $4d_0$, where d_0 is one atomic lattice spacing in [1 1 1] direction 0.2328 nm. The adjacent distance between the indenter

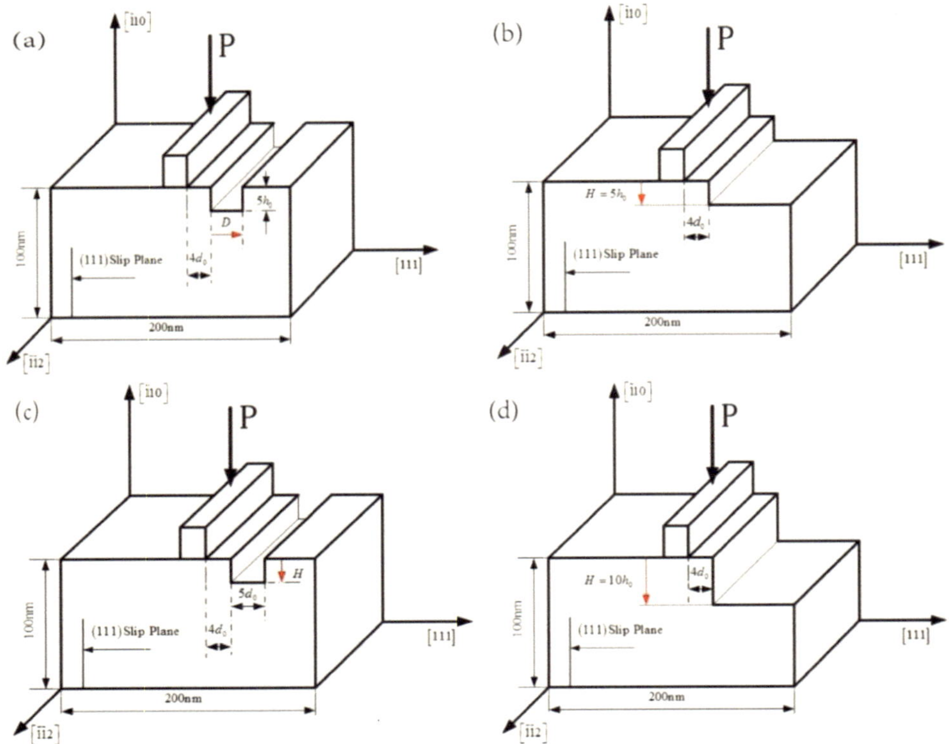

Figure 7.
Schematic representation of the nanoindentation model of size effect: (a) width (D) changing from $1d_o$ *to* $10d_o$ *of surface pit defect with the fixed height =* $5h_o$*; (b) the comparison model of surface step with height =* $5h_o$*; (c) height (H) changing from* $1h_o$ *to* $10h_o$ *of surface pit defect with the fixed width =* $5d_o$*; (d) the comparison model of surface step with height =* $10h_o$*.*

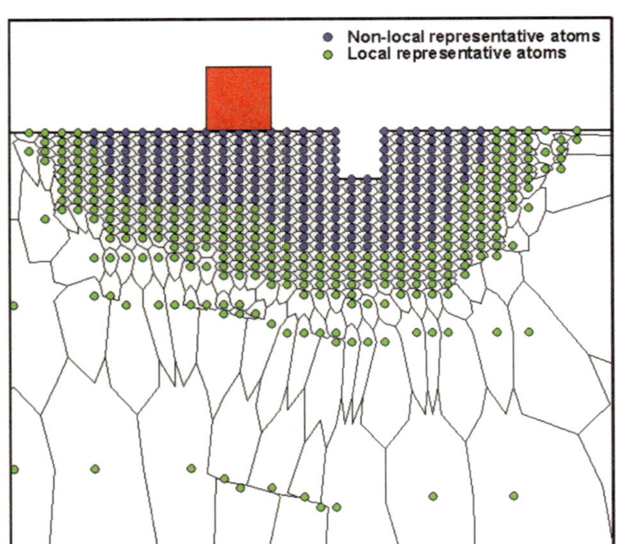

Figure 8.
Schematic representation of local and non-local representative atoms with initial surface pit defect.

and the pit in this model is chosen to be $4d_0$ (as shown in **Figure 7**), which is proved to be reasonable referring to the research on nanoindentation on a stepped surface [14], where the spatial extent of the step's influence has been figured out to be approximately three times the contact radius. Therefore, the contact radius of this simulation is $2d_0$ (half of the indenter width), and the distance between the left side of surface pit defect and the center of the indenter is exactly $6d_0$ ($4d_0+4d_0/2 = 6d_0$), which quite agrees with this reference [14] ($6d_0/2d_0 = 3$). Although the indentation tip shape is different between this simulation and the nanoindentation experiment, where it is square and round-like, respectively, actually, the nanohardness are both calculated by the real contact radius. That is to say, the change of the nanohardness in these two cases is quite similar if considering the ratio of defect distance to the real contact radius (as well as the experiment discussion). The simulation models of the width effect and height effect of surface pit defect are shown as **Figure 7**, and relevant parameter values are shown in **Table 3** in order to make a more comprehensive investigation. Besides, the comparison model of nanoindentation on the stepped surface step with the height of the step $10h_0$ has also been carried out as shown in **Figure 7(d)**. Further, this model keeps the boundary condition rigid at the bottom and free at the sides, and the thickness is equal to the minimal repeat distance. The displacement-imposed boundary condition forces the atoms under the indenter to move into the material gradually. Each load step of indentation has set 0.02 nm in order to be more proper and effective to catch the dislocation nucleation and mission. Moreover, the final depth is 1.2 nm, which ensures that the behavior in the vicinity of the indenter will not be affected by the far-field boundary conditions.

3.2 Results and discussion

3.2.1 Width effect on the yield load due to the pit defect

As is known to all that the yield load of materials is one of the most important factors of the material properties, however it can be obviously affected by defects such as surface pit defect. Normally, the yield load can be easily obtained from the first peak load in the load-displacement curve, which suggests onset of the elastic-to-plastic transition. In this chapter, we have taken 10 different widths of the pit from D = $1d_0$ to $10d_0$ with a fixed height H = $5h_0$, in order to investigate the width

Width effect		Height effect	
Width of pit (d_0)	**Height of pit (h_0)**	**Width of pit (d_0)**	**Height of pit (h_0)**
1	5	5	1
2			2
3			3
4			4
5			5
6			6
7			7
8			8
9			9
10			10

Table 3.
The models parameters of the width effect and height effect.

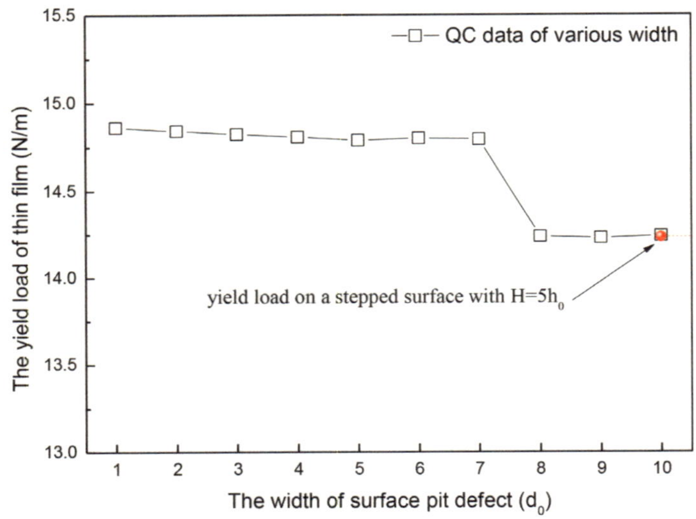

Figure 9.
The yield load of thin film as the width changing of surface pit defect (with a standard deviation of 0.01 N/m). QC—quasicontinuum method.

effect of surface pit defect on the yield load. The change of the yield load curve as pit width has been revealed in **Figure 9**. Generally, a reduction tendency of the yield load of Al thin film with the pit defect displays result from more and more serious destruction to the atomic structure by the increase of the pit width. Further, the yield load experiences an extremely slow reduction when the pit width increases from $D = 1d_0$ to $7d_0$; after that it obviously drops from 14.8 to 14.24 N/m when the pit width reaches $7d_0$. Then, the yield load decreases slowly again.

It is necessary and significant to compare the nanoindentation on a stepped surface with $H = 5h_0$ as shown in **Figure 7(b)**, where the pit width can be treated as infinitely large. The result is approximately 14.23 N/m, very close to the yield load in the case of $D = 10d_0$ (the red point in **Figure 9**). This implies that the yield load of Al thin film nearly equals the yield load value in the case of stepped surface when the pit width increases to $10d_0$.

3.2.2 Height effect on the yield load due to the pit defect

We have taken 10 different heights of the pit from $H = 1h_0$ to $10h_0$, with a fixed width $D = 5d_0$, in order to investigate the height effect of surface pit defect on the yield load as shown in **Figure 7(c)**. The change of the yield load curve as pit height has been revealed in **Figure 10**. Similarly, the yield load experiences an extremely slow reduction when the pit height increases from $H = 1h_0$ to $5h_0$; after that it obviously drops from 14.79 to 14.14 N/m when the pit width reaches $6d_0$. Then, the yield load decreases slowly again.

In the same way, it is also necessary and significant to compare the nanoindentation on a stepped surface with $H = 10h_0$ as shown in **Figure 7(d)**. The result is approximately 13.75 N/m (the red point in **Figure 10**), which is quite near the yield load 13.93 N/m in the case of $H = 10h_0$.

3.2.3 The investigation of dislocation nucleation and the estimation of Peierls stress

We have carried out a further probe of atomic snapshot and corresponding out-of-plane displacement plot, in order to explain the obvious drops of yield load

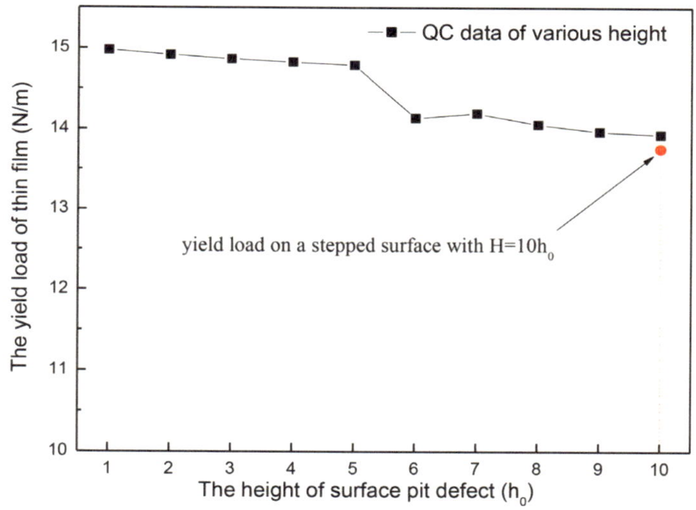

Figure 10.
The yield load of thin film as the height changing of surface pit defect (with a standard deviation of 0.01 N/m).

(D = $7d_0$ to $8d_0$ segment in **Figure 9**, H = $5h_0$ to $6h_0$ segment in **Figure 10**). Taking the cases of D = $1d_0$ in width effect simulation and H = $1h_0$ in height effect simulation, for example, it can be easily found out from nucleated dislocations and UZ contours displayed in **Figure 11** that two dissociated <1 1 0> edge dislocations are emitted beneath the indenter after nucleation during the thin film yielding. Moreover, the dislocated structure along with the out-of-plane displacements experienced by the atoms has also been displayed in **Figure 11**, with the dimension 0.1 nm, where a fingerprint of the dislocations has been clearly shown between the partials in the stacking fault regions. According to the structure of FCC metal, it can be easily found out that the dislocations are composed of 1/6 <1 1 2> Shockley partials. On the left,

$$\frac{1}{2}\left[\bar{1}10\right] = \underbrace{\frac{1}{6}\left[\bar{1}2\bar{1}\right]}_{top} + \underbrace{\frac{1}{6}\left[\bar{2}11\right]}_{bottom} \tag{5}$$

and on the right,

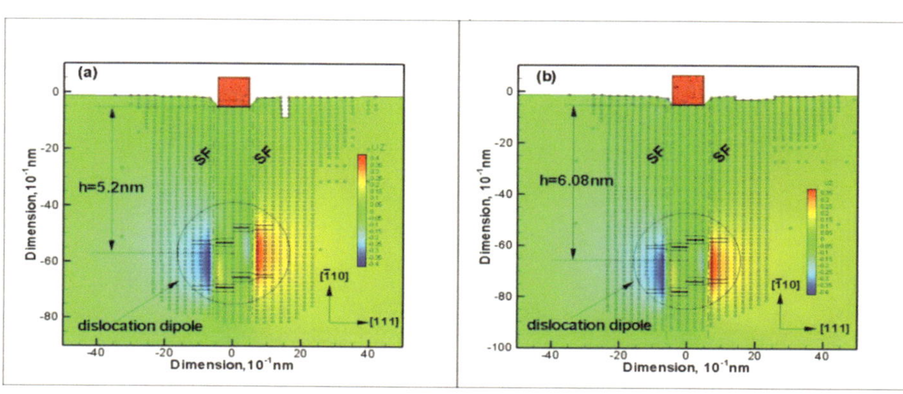

Figure 11.
Snapshot of atoms under the indenter and corresponding out-of-plane displacement plot, where UZ is atom displacement at out-of-plane: (a) width changing D = $1d_o$ at the yield of thin film; (b) height changing H = $1h_o$ at the yield of thin film.

$$\frac{1}{2}[1\bar{1}0] = \underbrace{\frac{1}{6}[1\bar{2}1]}_{\text{top}} + \underbrace{\frac{1}{6}[2\bar{1}\bar{1}]}_{\text{bottom}} \tag{6}$$

Comparing all these width and height effect cases in this simulation, we find that emission depth of dislocations changes due to the size of the pit. Take **Figure 11(a)** and **(b)**, for example, the dislocation dipole similarly travels into bulk after nucleation at the load step of 0.5 nm; however, its center of the emission depth settles, respectively, at the depth of 5.2 and 6.08 nm. It might be predicted that the yield load of thin film in macroscopy corresponds to the emission depth of dislocation in microscopy.

Considering that Peierls stress is exactly the resisting force during the dislocation movement due to the lattice structure, all these emission depths of dislocations have been adopted as an equilibrium distance to further calculate the Peierls stress predicted by the EAM potential [38]. Except the lattice friction, there are two forces acting on the dislocation: (i) the Peach-Koehler force (F_{PK}) due to the indenter stress field driving the dislocation into bulk and (ii) the image force (F_I) pulling the dislocation up to the surface. The dislocation, which is forced by the sum of these two forces, escapes the attractive region and propagates into the bulk and is finally stopped by lattice friction. Consequently, the force on the dislocation will be balanced at the equilibrium depth by the lattice friction force that is due to the Peierls stress (σ_P) [31].

$$F_{PK} + F_I = b\sigma_P \tag{7}$$

Shear stress beneath the indenter is necessary to be further obtained to calculate the Peach-Koehler force. In this simulation, the rectangular indenter is frictionless, applying to an elastic thin film occupying the lower half-plane. When $y < 0$, the shear stress in bipolar coordinates is [39]

$$\sigma_{xy} = -\frac{Pr^2 \sin\theta}{\pi(r_1 r_2)^{3/2}} \sin\left[\theta - \frac{3}{2}(\theta_1 + \theta_2)\right] \tag{8}$$

where P is the indentation load. According to the coordinate system of $2a$ indentation contact (the width of indenter is $2a$), as shown in **Figure 12**, at a depth h beneath the right indenter tip, there is $r = \sqrt{a^2 + h^2}$, $r_1 = h$, $r_2 = \sqrt{4a^2 + h^2}$, $\theta = -\tan^{-1} h/a$, $\theta_1 = -\pi/2$, and $\theta_2 = -\tan^{-1}(h/2a)$. The Peach-Koehler force is

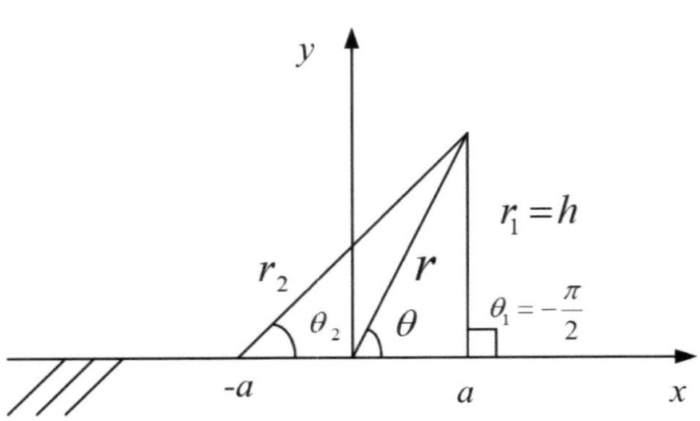

Figure 12.
Bipolar coordinate for a 2a *indentation contact.*

$$F_{\mathrm{PK}}(h) = (\mathbf{b} \cdot \boldsymbol{\sigma}) \times \ell = b\sigma_{\mathrm{xy}}(h) \tag{9}$$

where $\boldsymbol{\sigma}$ is the applied stress tensor, $\boldsymbol{b}$ is the Burgers vector, and ℓ is the dislocation line vector.

The image force applying to the dislocation dipole with width $d = 2a$ at the emission depth h can be displayed as follows:

$$F_{\mathrm{I}} = \frac{\mu b^2}{\pi(1-v)}\left[\frac{1}{4h} - \frac{4h^3(4h^2 - 3d^2)}{(4h^2 + d^2)^3}\right] \tag{10}$$

Peierls stress of all these cases of pit size has been calculated and plotted based on the discussion above, where **Figures 13** and **14**, respectively, show Peierls stress of width effect and height effect. In the case of width effect, Peierls stress maintains

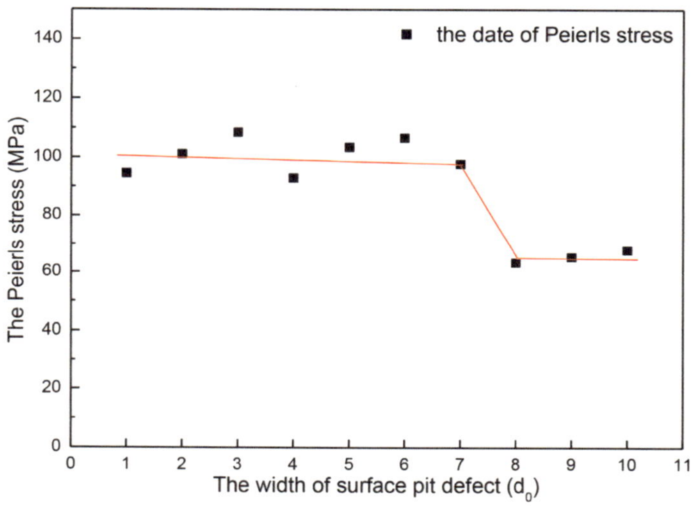

Figure 13.
The variation of Peierls stress in the simulation of width effect (with a standard deviation of 0.2 MPa).

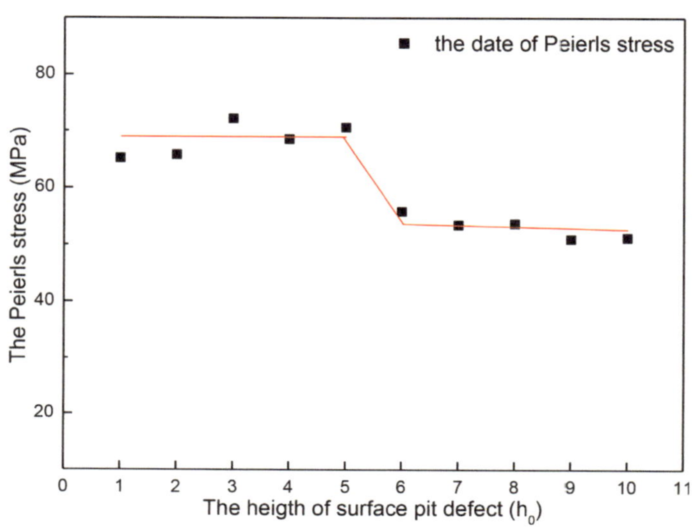

Figure 14.
The variation of Peierls stress in the simulation of height effect (with a standard deviation of 0.1 MPa).

around the value of 100 MPa with tiny fluctuation from D = $1d_0$ to $7d_0$; after that the Peierls stress displays a sudden obvious dropdown to about 70 MPa, which is quite similar compared with yield load curve in **Figure 9**. In the case of height effect, Peierls stress maintains around the value of 70 MPa with tiny fluctuation from H = $1h_0$ to $5h_0$; after that the Peierls stress displays a sudden obvious dropdown to about 50 MPa, which is also greatly in accordance with the yield load curve in **Figure 10**. The conclusion can be drawn that such obvious decline of yield load (D = $7d_0$ to $8d_0$ segment in **Figure 9**, H = $5h_0$ to $6h_0$ segment in **Figure 10**) is closely related to the severe reduction of the Peierls stress, suggesting that it is reasonable and effective to explain the variation of yield load through the Peierls stress.

3.2.4 Size coefficient

We make a further probe on the difference of turning point between width effect and height effect, corresponding to D = $7d_0$ in the width effect simulation and H = $5h_0$ in the height effect simulation. It can be predicted that the influence degree of width factor is different from the height factor. It is necessary to quantify the size effect of surface pit defect to explain the reason of these differences. It can also be easily recognized that if the distance between the pit and the indenter decreases, the influence would be much more severe on the hardness and yield load. Namely, controlling the same influence of the pit on the nanohardness, the larger size of the pit is required when the pit goes far away from the indenter. Therefore, we defined a size coefficient α as the following, which is dimensionless in order to explain the size effect of surface pit defect:

$$\alpha = \frac{L^*}{d^*} \tag{11}$$

where "L^*" is the characteristic length of the pit, such as D in the width effect simulation or H in the height effect simulation and "d^*" is the distance between the center of the indenter and the left boundary of the pit, namely, $6d_0$ in this simulation.

According to the simulation result of width effect, the critical width value to make a sudden obvious drop of yield load is $7d_0$ (as shown the point D = $7d_0$ in **Figure 9**). Consequently, the size coefficient α is approximately 1.17 ($\frac{L^*}{d^*} = \frac{D}{d^*} = \frac{7d_0}{6d_0} = \frac{7}{6}$). When α reaches approximately 2 ($\frac{L^*}{d^*} = \frac{D}{d^*} = \frac{10d_0}{6d_0} = 1.7$), as shown in the point D = $10d_0$ in **Figure 9**, the yield load of thin film with surface pit defect nearly equals the one of nanoindentation with surface step as shown in the red point in **Figure 9**.

According to the simulation result of height effect, the critical height value to make a sudden obvious drop of yield load is $5h_0$ (as shown the point H = $5h_0$ in **Figure 10**). Consequently, the size coefficient α is approximately 0.51 ($\frac{L^*}{d^*} = \frac{H}{d^*} = \frac{5h_0}{6d_0} = 0.51$). When α reaches approximately 1 ($\frac{L^*}{d^*} = \frac{H}{d^*} = \frac{10h_0}{6d_0} = 1.02$), as shown in the point H = $10h_0$ in **Figure 10**, the yield load of thin film with surface pit defect nearly equals the one of nanoindentation with surface step as shown in the red point in **Figure 10**.

By contrast, the size coefficient of height is approximately half of the one of width to boost the sudden decline of yield load, implying that the height of the pit has a greater influence on the yield load than the width.

Moreover, the change of yield load of thin film as the pit area has been plotted and shown in **Figure 15**, where we can easily find out that the slope of yield load curve by height increasing is larger than the one by width increasing. That is to say, the height increasing makes the yield load decrease faster. Besides, the yield load by

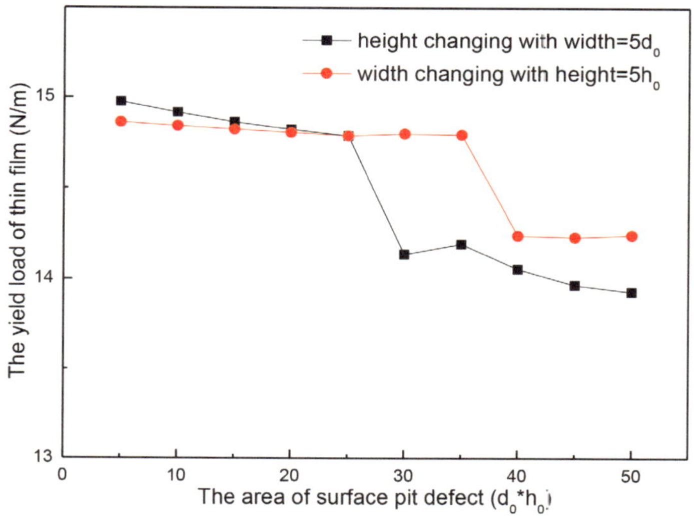

Figure 15.
The yield load of thin film as the area changing of surface pit defect.

width increasing is smaller than the one by height increasing during $5h_0d_0$ to $25h_0d_0$ segment of the pit area, which results from the height of the pit in width increase curve (as shown red curve in **Figure 15**) is larger than the other one (black curve). If the area increases over than $25h_0d_0$, the yield load by height increase is smaller than the one by the width increase. It can be well explained that the height of the pit in the curve of height increase goes up over $6h_0$, while the one in the curve of width increase still maintains $5h_0$, suggesting that the height of the pit has played a more important role on yield load.

4. Conclusions

Through this whole chapter, we apply the QC method to study the influence of surface pit defect in nanoindentation. Fifteen distances of adjacent boundaries between the pit and indenter have been taken into account to study the distance effect, compared with the nanoindentation on defect-free surfaces, while various sizes of the pit have been taken into account to study the size effect, compared with the nanoindentation on the stepped surfaces. Based on the discussion, we can make some conclusions as the following:

i. Compared with the nanoindentation on the defect-free surface, the pit defect plays a significant role in the elastic-to-plastic transition that leads to the decrease of the nanohardness.

ii. The nanohardness increases in a wave that goes up in a period of three atoms when the pit moves far away from the indenter atom by atom, which is strongly linked to the crystal structure of periodic atom arrangement on {1 1 1} atomic close-packed planes of FCC metal. Moreover, it can be nearly considered as no influence on the nanohardness if the adjacent distance between the pit and indenter increases over than 16 atomic spacing.

iii. The formula for critical load of dislocation emission of Al thin film has been effectively revised with initial pit defect, where the reduction trend of nanohardness, as the increasing distance between the pit and the indenter, is in great accordance with the experimental results of nanoindentation on the stepped surface with various distances. Such study might be referential to the research of material properties with defects, especially in microchips and MEMS.

iv. When the size coefficient of width (of height) equals about one unit (half unit), the yield load experiences an obvious drop. When it reaches about two units (one unit), the yield load is nearly close to that of the nanoindentation on a stepped surface.

v. Compared to the width, the height of surface pit defect shows a greater influence on the yield load of thin film, implying that the height of the pit is a leading factor on the influence of yield load, which might have great significance to the defect design and applications in artificial materials.

Acknowledgements

The authors would like to thank Professor Tadmor E. B along with his cooperative partners for their open source QC code. Besides, this work is supported by the National Natural Science Foundation of China (Grant No. 11572090).

Conflict of interest

The authors declare that the founding sponsors had no contributions in the design of the study; in the collection, analyses, or interpretation of data; in the writing of the manuscript; and in the decision to publish the results. So, there is no conflict of interest.

Author details

Zhongli Zhang[1,2], Yushan Ni[1*], Jinming Zhang[2], Can Wang[2] and Xuedi Ren[2]

1 Department of Aeronautics and Astronautics, Fudan University, Shanghai, China

2 Shanghai Institute of Measurement and Testing Technology, Shanghai, China

*Address all correspondence to: niyushan@fudan.edu.cn

References

[1] Oliver WC, Pharr GM. An improved technique for determining hardness and elastic modulus using load and displacement sensing indentation experiments. Journal of Materials Research and Technology. 1992;7:1564-1583

[2] Li XD, Bhushan B. A review of nanoindentation continuous stiffness measurement technique and its applications. Materials Characterization. 2002;**48**:11-36

[3] Bamber MJ, Cooke KE, Mann AB, Derby B. Accurate determination of Young's modulus and Poisson's ratio of thin films by a combination of acoustic microscopy and nanoindentation. Thin Solid Films. 2001;**399**:299-305

[4] Zhu PZ, Hu YZ, Fang FZ, Wang H. Multiscale simulations of nanoindentation and nanoscratch of single crystal copper. Applied Surface Science. 2012;**258**:4624-4631

[5] Chen J, Bull SJ. Assessment of the toughness of thin coatings using nanoindentation under displacement control. Thin Solid Films. 2006;**494**:1-7

[6] Sangwal K, Gorostiza P, Sanz F. Atomic force microscopy study of nanoindentation creep on the (100) face of MgO single crystals. Surface Science. 2000;**446**:314-322

[7] Zhang TH, Yang YM. The application of nanohardness technology in the mechanical properties testing of surface engineering. Chinese Journal of Mechanical Engineering. 2002;**24**:85-88

[8] Mitchell TE. Dislocations and plasticity in single crystals of face centered cubic metals and alloys. Progress in Applied Materials Research. 1964;**6**:117-238

[9] Mitchell JW. In: Doremus RH, Roberts BW, Turnbull D, editors. Growth and Perfection of Crystals. New York, NY, USA: Wiley; 1958. pp. 386-389

[10] Yang B, Vehoff H. Dependence of nanohardness upon indentation size and grain size—A local examination of the interaction between dislocations and grain boundaries. Acta Materialia. 2007; **55**:849-856

[11] Soifer YM, Verdyan A, Kazakevich M, Rabkin E. Nanohardness of copper in the vicinity of grain boundaries. Scripta Materialia. 2002;**47**:799-804

[12] Yu WS, Shen SP. Multiscale analysis of the effects of nanocavity on nanoindentation. Computational Materials Science. 2009;**46**:425-430

[13] Shan D, Yuan L, Guo B. Multiscale simulation of surface step effects on nanoindentation. Materials Science and Engineering A. 2005;**412**:264-270

[14] Keily JD, Hwang RQ, Houston JE. Effect of surface steps on the plastic threshold in nanoindentation. Physical Review Letters. 1998;**81**:4424-4427

[15] Gouldstone A, Van Vliet KJ, Suresh S. Nanoindentation: Simulation of defect nucleation in a crystal. Nature. 2001;**411**:656-657

[16] Zimmerman JA, Kelchner CL, Klein PA, Hamilton JC, Foiles SM. Surface step effects on nanoindentation. Physical Review Letters. 2001;**87**: 165507-165561

[17] Jiang WG, Su JJ, Feng XQ. Effect of surface roughness on nanoindentation test of thin films. Engineering Fracture Mechanics. 2008;**75**:4965-4972

[18] Miller RE, Shilkrot LE, Curtin WA. A coupled atomistics and discrete dislocation plasticity simulation of

nanoindentation into single crystal thin films. Acta Materialia. 2004;**52**:271-284

[19] Tsuru T, Shibutani Y. Anisotropic effects in elastic and incipient plastic deformation under (001), (110), and (111) nanoindentation of Al and Cu. Physical Review B: Condensed Matter and Materials Physics. 2007;**75**: 035415-035421

[20] Wagner RJ, Ma L, Tavazza F, Levine LE. Dislocation nucleation during nanoindentation of aluminum. Journal of Applied Physics. 2008;**104**: 114311-114312

[21] Li JW, Ni YS, Lin YH, Luo C. Multiscale simulation of nanoindentation on Al thin film. Acta Metallurgica Sinica. 2009;**45**:129-136

[22] Zhang ZL, Ni YS. Multiscale analysis of delay effect of dislocation nucleation with surface pit defect in nanoindentation. Computational Materials Science. 2012;**62**:203-209

[23] Zhang Z, Ni Y, Zhang J, Wang C, Jiang K, Ren X. Multiscale simulation of surface defects influence nanoindentation by a quasi-continuum method. Crystals. 2018;**8**:291-300

[24] Zhang ZL, Ni YS, Zhang JM, Wang C, Ren XD. Multiscale analysis of size effect of surface pit defect in nanoindentation. Micromachines. 2018; **9**:298-309

[25] Qin ZD, Wang HT, Ni YS. Multiscale simulations of FCC Al nanoindentation. Chin. Q. Mech. 2007;**1**: 46-53

[26] Tadmor EB. The Quasicontinuum method [PhD thesis]. Providence, RI, USA: Brown University; 1996

[27] Ercolessi F, Adams JB. Interatomic potentials from first-principles calculations: The force-matching method. Europhysics Letters. 1994;**26**: 583-588

[28] Tadmor EB, Ortiz M, Phillips R. Quasicontinuum analysis of defects in solids. Philosophical Magazine A. 1996; **73**:1529-1563

[29] Tadmor EB, Phillips R, Ortiz M. Mixed atomistic and continuum models of deformation in solids. Langmuir. 1996;**12**:4529-4534

[30] Shenoy VB, Miller R, Tadmor EB, Rodney D, Phillips R, Ortiz M. An adaptive finite element approach to atomic-scale mechanics—The quasicontinuum method. Mechanics and Physics of Solids. 1999;**47**:611-642

[31] Tadmor EB, Miller R, Phillips R. Nanoindentation and incipient plasticity. Journal of Materials Research and Technology. 1999;**14**:2233-2250

[32] QC Tutorial Guide Version 1.4. Available online: http://qcmethod.org/documentation [Accessed: 26 June 2018]

[33] Nanoindentation. Available online: https://en.wikipedia.org/wiki/Nanoindentation [Accessed: 2 May 2018]

[34] Jiang WG, Li JW, Su JJ, Tang JL. Quasicontinuum analysisi of indenter size effect in nanoindentation tests. Chinese Journal of Solid Mechanics. 2007;**4**:375-379

[35] Crystal Strucrure. Available online: https://en.wikipedia.org/wiki/Crystal_structure [Accessed: 20 June 2018]

[36] Lu G, Kioussis N, Bulatov VV, Kaxiras E. Generalized-stacking-fault energy surface and dislocation properties of aluminum. Physical Review B. 2000;**62**:3099-3108

[37] Abu Al-Rub RK, Voyiadjis GZ. A physically based gradient plasticity theory. International Journal of Plasticity. 2006;**22**:654-684

[38] Daw MS, Baskes MI. Semiempirical, quantum mechanical calculation of hydrogen embrittlement in metals. Physical Review Letters. 1983;**50**:1285

[39] Muskhelishvili NI. Some Basic Problems of the Mathematical Theory of Elasticity. 3rd ed. Groningen, The Netherlands: P. Noordhoff Ltd.; 1953. pp. 481-483